Samara S. Vieira Alves
José Leôncio de A. Silva
José F. de Medeiros

Crop performance under salt stress

Samara S. Vieira Alves
José Leôncio de A. Silva
José F. de Medeiros

Crop performance under salt stress

Studies in representative soils of the Mossoró-Assu agropolis

ScienciaScripts

Imprint
Any brand names and product names mentioned in this book are subject to trademark, brand or patent protection and are trademarks or registered trademarks of their respective holders. The use of brand names, product names, common names, trade names, product descriptions etc. even without a particular marking in this work is in no way to be construed to mean that such names may be regarded as unrestricted in respect of trademark and brand protection legislation and could thus be used by anyone.

Cover image: www.ingimage.com

This book is a translation from the original published under ISBN 978-3-330-77203-8.

Publisher:
Sciencia Scripts
is a trademark of
Dodo Books Indian Ocean Ltd. and OmniScriptum S.R.L publishing group

120 High Road, East Finchley, London, N2 9ED, United Kingdom
Str. Armeneasca 28/1, office 1, Chisinau MD-2012, Republic of Moldova, Europe
Managing Directors: Ieva Konstantinova, Victoria Ursu
info@omniscriptum.com

Printed at: see last page
ISBN: 978-620-8-64068-2

SUMMARY

1 INTRODUCTION

The salinization of soils can result from the material of origin and/or inadequate management of agricultural techniques, such as excessive fertilization and irrigation with water unsuitable for this purpose, transforming fertile and productive land into arid and unsuitable for agriculture.

The Mossoró-Assu Agropólo is the main fruit, vegetable and grain producing region in the semi-arid region of the state of Rio Grande do Norte, where rain-fed and mainly irrigated agriculture is practiced. This region is bathed by two of the largest hydrographic basins in the state of Rio Grande do Norte (the Apodi-Mossoró River and the Piranhas-Açu River) and also has large volumes of groundwater in the calcareous aquifers (medium and high salinity water) and sandstone aquifers (low salinity water), mainly in the Chapada do Apodi.

This abundance of water allows irrigation to be practiced, but in some irrigated areas there have been problems with soil salinization and loss of crop yield associated with inadequate soil, water and plant management. Although these problems are more evident in clay soils that are not very well drained and have high clay activity, until now these particularities have not been properly considered by farmers, technicians and some researchers in the region. Generally, any type of water and the same management is used, regardless of the type of soil and crop.

The source materials for the soils of the Chapada do Apodi are sedimentary in nature, composed mainly of limestone, sandstone, barrier group sediments and alluvial sediments. This leads to the formation of soils with a wide range of chemical, physical, mineralogical and morphological characteristics. The most common soils in the region are Cambissolos, Neossolos Litólicos, Chernossolos Rêndzicos, Neossolos Flúvicos, Vertissolos, Argissolos and Latossolos. Considering the above, it is believed that the texture and mineralogy of the soils are characteristics that should be taken into account when establishing appropriate water and plant management in irrigated agriculture in the region. However, there is little research in the literature showing the influence of salinity on different types of soil.

Salinity problems are more frequent in tropical regions with hot, dry climates, characterized by high evapotranspiration rates and low rainfall, such as the Semiarid region of northeastern Brazil, where salinization results from the physical and chemical nature of the soils, the rainfall regime and high evaporation. In addition, the use of irrigation leads to the incorporation of salts into the soil profile, due to the soluble ions present in the water which, with constant use without leaching, causes the salt to be deposited in the root system zone due to evaporation. Restricted drainage also contributes to the salinization of irrigable soils and can influence the height of the water table. Soil salinization progresses when the amount of salts accumulated by irrigation water is greater than the amount removed by drainage water (PIZARRO, 1985; AYERS; WESTCOT, 1999).

According to Sumner (1995), many soils found in semi-arid climates have appreciable quantities of weatherable minerals (feldspars, hornblende, plagioclase, calcite and gypsum), which can maintain high activities of calcium, magnesium and sodium ions in solution. In addition to these factors, there is also the high evapotranspiration rate and the presence of underground aquifers with water enriched in salts (RICHARDS, 1954; SZABOLCS, 1995). Therefore, the identification of mineralogy is important, as some soils have elements in their constitution that can contribute to the salinization and, or, alkalinization of the soil, which is of fundamental importance for understanding this process that has a close relationship with agricultural sustainability.

Silva Junior *et al.* (1999), studying the ionic composition of crystalline water in the northeast of Brazil (states of Paraiba, Rio Grande do Norte and Cearà), found that there is a predominance of sodic chloride water, with some variation, regardless of the level of salinity, locations and sources of origin. Morais *et al.* (1998), evaluating the quality of irrigation water in analytical samples from the database of the Soils and Geology Department of ESAM (Mossoró Higher School of Agriculture), found that 47% and 19% of the samples were restricted for agricultural use , slight to moderate and severe restriction, respectively.

Excess salts in the soil reduce the availability of water for plants, as well as causing problems due to the effect of specific toxic ions, affecting the yield and quality of crops.

The melon, bean, corn, sorghum and pepper are crops that require high night and day temperatures throughout the cycle, as high temperatures and low relative humidity considerably improve the quality of the fruit, making it more consistent and with a longer post-harvest shelf life. These crops are widely grown in the state of Rio Grande do Norte, with melon being of great economic importance for export and the others for family farming and production for domestic consumption.

Mixing water with a high concentration of salts with good quality water, although not a common practice in the semi-arid northeast, has been an alternative for improving the quality and increasing the availability of water in these regions, especially in areas where the water has a high salt content. Thus, in addition to increasing the volume of water available per area, it reduces the RAS and EC values of diluted water, directly influencing the reduction of the risk of sodification and salinization in soils and, consequently, the percentage of seedling emergence and plant development. In this case, the use of these waters is dependent on the tolerance of the crops to salinity and the management of irrigation with a view to controlling the salinization of these areas.

2. OBJECTIVES

2.1 GENERAL OBJECTIVE

To evaluate the effects of irrigation water salinity on the soil and on the performance of melon, cowpea, corn, sorghum and pepper plants grown in soils of the Mossoró-Assu Agropolis.

2.2 SPECIFIC OBJECTIVES

a) To study the evolution of soil salinity when irrigated with water of different saline levels;

b) Evaluate the influence of salinity in different types of soil on the behavior of different crops; and

c) To evaluate the effect of salinity and soil on the nutrition of melon, bean, corn, sorghum and pepper plants.

3. THEORETICAL FRAMEWORK

3.1 AGROPÓLO MOSSORÓ-ASSU

Agropólo Mossoró-Assu is the main fruit, vegetable and grain producing region in the semi-arid region of the state of Rio Grande do Norte, covering part or all of the area of 18 municipalities, where rain-fed and mainly irrigated agriculture is practiced. This region is bathed by two of the largest hydrographic basins in the state of Rio Grande do Norte (the Apodi- Mossoró River and the Piranhas-Açu River) and also has large volumes of groundwater in the calcareous aquifers (medium and high salinity water) around 100 m deep and the arenito-açu aquifer (low salinity water), with an average depth of 1000 m. According to Medeiros, Lisboa and Oliveira (2003) the types of wells most used in the region are those that exploit the calcareous aquifer, due to the low cost of investment, maintenance and abstraction, but these waters have a relatively high concentration of salts, which can exceed 2000 mg L^{-1} (EC= 3.0 dS $m^{-1)}$. In addition to the lower cost of abstraction, the advantage of Jandaira limestone water is that it quickly replenishes the groundwater table during the rainy season, as it is closer to the surface, allowing it to receive an average annual recharge equivalent to a flow of $160x10^6$ m^3 $year^{(-1)}$, However, this depends on the rainfall regime during the period, while in the Açu sandstone the recharge is only $5x10^6$ m3 $year^{(-1})$ (REBOUÇAS, 1967), and due to its quality it is also used for human consumption. In this case, the use of this saline water source depends on the tolerance of the crops to salinity, the irrigation system adopted and the management used, which should aim to monitor and control soil salinization.

The main factor that has led to the use of water from the limestone aquifer is the lower cost of drilling, which is carried out by small local companies, with maintenance and pumping costs that are accessible to small producers. On the other hand, wells in the Açu Sandstone are generally drilled by multinational oil companies, costing around 20 times the amount of drilling in the Limestone aquifer.

The great abundance of water in the region allows irrigation to be practiced, but in some irrigated areas there have been problems of soil salinization and loss of crop yield associated with inadequate soil, water and plant management. Although these problems are more evident in clay soils, with restricted drainage and high clay activity, until now these particularities have not been properly considered by farmers, technicians and some researchers in the region. Generally, any type of water and the same water management is used, regardless of the type of soil and crop.

The source materials for the soils of the Chapada do Apodi are sedimentary in nature, composed mainly of limestone, sandstone, barrier group sediments and alluvial sediments (DIÓGENES, 2012). This leads to the formation of soils with a wide range of chemical, physical, mineralogical and morphological characteristics. The most common soils in the region are Cambisols, Litholic Neosols, Rhenitic Chernosols, Fluvial Neosols, Vertisols, Argisols and Latosols (BRASIL, 1971; ERNESTO SOBRINHO, 1980). It is believed that the texture and mineralogy of the soils are characteristics that should be taken into account when establishing appropriate water and plant management in irrigated agriculture in the region.

Despite the availability of technology in the Mossoró-Assu agropolis region, with the use of excellent quality irrigation equipment, generally imported from Israel, it can be seen that irrigation projects have low uniformity, in some cases below 70%, increasing water consumption (PEREIRA, 2010). One aspect that could contribute to reducing water consumption in agriculture and making irrigation more technologically advanced would be the use of Kc calculations, leaching analysis, as well as direct and indirect monitoring of soil moisture.

3.2 WATER QUALITY FOR AGRICULTURE

The variation in the composition and quality of water used for irrigation depends on the climate zone, the source of the water, the route taken, the time of year, the geology of the region and the development of the irrigation (SHALHEVET; KAMBUROV, 1976). Generally, waters from humid zones have a lower salt content than those from arid zones; underground waters are saltier than those from rivers and the latter, near the mouth, contain more salts than those near the source; the concentration of salts in river waters varies with the seasons, being higher in the fall than in the spring (SOUSA,

2006).

The semi-arid region has a number of areas that are prone to salt accumulation processes. Existing water sources are an important input for agriculture in the region, but their quality varies over time and space. During the dry season, water from wells, dams and other sources has a high saline concentration. Due to the high temperatures, crop evapotranspiration is higher, requiring a greater supply of water.

When assessing water quality for agriculture, three important parameters must be taken into account: salinity, sodicity and ion toxicity. The effect of salinity is osmotic in nature and can directly affect crop yields. Sodicity refers to the relative effect of sodium in irrigation water, which tends to increase the percentage of exchangeable sodium in the soil (PST), damaging its physical and chemical properties and causing infiltration problems. High PST values, especially in low salinity conditions, cause particle dispersion, with a reduction in the soil's hydraulic conductivity. The sodium adsorption ratio (SAR) of the irrigation water is the parameter with the best correlation with the PST of the soil. Some ions are important parameters because of the toxicity they can cause to plants. Chloride, sodium and boron ions, when present in high concentrations in water, can cause damage to crops, affecting their production (EMBRAPA, 2010).

In many places, good quality water may not be enough to maintain irrigated agriculture, or its cost is high, such as the water from the Açu Sandstone, as discussed in the previous topic. On the other hand, most of the waters of the Chapada do Apodi are classified as saline and with little risk of sodicity (MAIA *et al.*, 1997).

The intensive use of brackish water, associated with inadequate fertilizer management, such as excessive application of fertilizers with a high saline content, can lead to an accumulation of salts in the root zone, and in some cases it is necessary to remove part of the salts through leaching to avoid salinization of the soil, which when in excess have deleterious effects on plant growth, (HOLANDA, 2010).

The main and most recommended classification of water for irrigation is based on three criteria: the first, using the EC of the water, is related to the risk of salinization; the second criterion is related to the risk of sodicity or problems with the structure of aggregates and water infiltration into the soil; and the third refers to the risk of toxicity by specific ions (AYERS; WESTCOT, 1999).

Until now, there has been no proper classification of water for Brazilian conditions, taking into account the water, soil, plant and climate, and its applicability to the different technological levels applied in relation to irrigation management, The most widely used classification is that proposed by the UCCC (University of California Committee of Consultants), cited by Ayers and Westcot (1999) and Pizzaro (1985) and that of the Salinity Laboratory in the United States (RICHARDS, 1954). For classifying the risk of infiltration and ionic toxicity, the guidelines presented by Ayers and Westcot (1999) are more appropriate.

Silva Junior *et al.* (1999) studied the ionic composition of crystalline water in the northeast of Brazil (states of Paraiba, Rio Grande do Norte and Cearà) and found that there is a predominance of sodic chloride waters, with some variation, regardless of the level of salinity, locations and sources of origin. Morais *et al.* (1998) evaluated the quality of irrigation water in analytical samples from the database of the Soils and Geology Department of ESAM (Mossoró Higher School of Agriculture) and found that 47% and 19% of the samples presented slight to moderate and severe restrictions for agricultural use, respectively.

In the melon-producing region located on the Apodi plateau, groundwater of calcareous origin (shallow wells) has high electrical conductivity, high levels of chlorides, low sodicity and high alkalinity (MEDEIROS, LISBOA and OLIVEIRA, 2003).

Around 70% of water sources in the Northeast are of good quality for irrigation with low or medium salinity (HOLANDA *et al.*, 2010), unlike the figures for the Apodi plateau, where around 60% of the sources pose a risk of salinization, however in regions that use good quality water, such as irrigated perimeters, it is estimated that 25 to 30% of these areas have some salinity problem, but these problems are related to inadequate management of the soil-water-plant system.

Mixing water with high salt concentrations with good quality water, although not a common practice in the Semi-Arid Northeast, has been an alternative for improving the quality and increasing the availability of water in these regions, especially in areas where the water has high salt contents. Thus, in addition to increasing the volume of water available per area, it reduces the RAS and EC values of diluted water, directly influencing the reduction of the risk of sodification and salinization in soils, directly influencing the percentage of seedling emergence and plant development.

Maia, Morais and Oliveira (2001), studying the ionic composition of water used for irrigation in the Mossoró-Assu agropolis, found that water from the Baixo Açu irrigated perimeter was classified as sodic, magnesian-sodic and sodic for the tube well, Amazon well and river sources, respectively, and the water from the Chapada do Apodi was classified as acidic-sodic, acidic and chloride for the tube well, Amazon well and river sources, respectively.

Silva Jùnior, Gheyi and Medeiros (1999), in a study on the chemical composition of crystalline waters in the Northeast of Brazil, observed a predominance of chloride-sodic waters, regardless of the salinity level and the locations of the sources of origin, and low levels of sulphate, carbonate and potassium. Chloride levels can reach values higher than 80% of the total anions present in waters with high salinity levels (ECa>3000 $\mu S\ cm^{-1}$), and waters with low salinity (ECa<750 $\mu S\ cm^{-1}$) can have bicarbonate concentrations very close to or higher than those of chloride (SILVA JÙNIOR, GHEYI and MEDEIROS, 1999).

Oliveira and Maia (1998), based on an analysis of 600 water samples from various aquifers of sedimentary origin in Rio Grande do Norte, showed that the highest incidence of salinity problems, expressed by their high electrical conductivity and sodicity, occurred in samples from wells that exploit the resources of the Jandaira limestone or from aquifers supplied directly by the water table. They also showed that in the sedimentary area of Rio Grande do Norte, saline waters with EC $\geq 3\ dS\ m^{-1}$ come proportionally from tube wells (27%), Amazon wells (cacimbas) (16%) and free springs (dams, rivers and streams) (21%).

3.3 SOIL SALINIZATION

The degradation of soils by salinization has become an increasingly frequent problem in arid regions, where low rainfall is not enough to promote the leaching and removal of salts. The predominance of sodium in relation to the other cations in the exchange complex can promote the dispersion and migration of colloids in the soil profile, clogging pores and reducing the movement of air and water, hindering plant growth (FREIRE *et al.*, 2003a).

The accumulation of salts in the soil depends on the quality of the irrigation water, the physical properties of the soil and, above all, the drainage conditions and the balance of water and salts in the subsoil (DIAS *et al.*, 2005).

Restricted drainage is a factor that generally contributes to soil salinization and may involve the presence of high amounts of groundwater or low soil permeability.

A soil is considered saline when its electrical conductivity in the saturation extract is greater than 4 dS m^{-1}, and is easily recognized by the presence of white crusts on the surface. The chemical characteristics of soils classified as saline are determined mainly by the types and quantities of salts present (RHOADES; KANDIAH; MASHALI, 1992).

The suitability of soils for agriculture depends largely on their conductivity to water and air and their properties related to fertility control. Permeability and inadequate hydraulic conductivity are the main problems in irrigated areas, because, unlike saline soils, sodic soils can have greatly reduced permeability and be difficult to prepare, because certain physico-chemical reactions associated largely with the colloidal fraction of the soil, are manifested mainly by the destructuring of aggregates and dispersion of clay minerals (RHOADES; KANDIAH; MASHALI, 1992).

In order to understand how the weak physical properties of sodic soils develop, it is necessary to look at the linked mechanisms involving the negatively charged colloidal clays of the soil, the organic matter of the soil and the layer of adsorbed cations electrostatically associated around the colloids and how exchangeable sodium, electrolyte concentration and pH affect this association. The cations are subject to two opposing processes: a) they are attracted by the electrostatic forces on the surfaces of the clay and the negatively charged organic matter; b) they tend to diffuse away from these surfaces, where their concentration is high, into the volume of the solution where their concentration is generally lower. Both processes result in an approximately exponential drop in the concentration of the cations from the clay surfaces to the volume of the solution. Divalent cations, such as calcium, are attracted to the negatively charged surface twice as strongly as monovalent cations such as sodium. In this way, the layer is also compressed by an increase in the electrolyte concentration of the solution, since the tendency of the cation to diffuse away decreases with a reduction in the concentration gradient (RHOADES; KANDIAH; MASHALI, 1992).

Soil aggregates are reduced when the cation layer expands (relative to the surface of the particle) and increase when it is compressed. Electrostatic charges of the same sign, which repel each other, and electrostatic charges of opposite signs, which attract each other, have relatively far-reaching effects. On the other hand, the adhesive forces, called van der Wall forces, and the chemical bonds involved in the interparticle associations that unite such units into microaggregates, are short-range forces. The greater the compression of the cation layer against the surface of the particle, the less the layers overlap and the less the repulsion between adjacent particles for a given distance. Consequently, the particles can approach each other more closely, to the point where adhesive forces predominate and aggregates are formed (RHOADES; KANDIAH; MASHALI, 1992).

The phenomenon of repulsion between the particles causes the soil solution to soak between them (dispersion). Because the clay particles are laminar in shape and tend to be arranged in parallel orientations, dispersion reduces the size of the pore space between the aggregates and therefore the permeability. Dispersion is primarily important in soils containing quantities of phyllosilicate clay mineral substances with expansive layers (smectites and montmorillonites) in which the PST exceeds around 15%. The reason for this is that, in these minerals, the sodium ions in the soil solution are first attracted to the outer surface of the clay plates. Only after the external charges have been completely neutralized do the sodium ions occupy the space between the parallel platelets of the oriented and associated clay particles of the microaggregates, where they create the repulsion forces between adjacent platelets that caused the expansion (RHOADES; KANDIAH; MASHALI, 1992).

Dispersion and disintegration of the soil can occur even in conditions of low PST values (<15), provided that the electrolyte concentration is sufficiently low. The assembly of aggregates results in more pore space than individual particles or microaggregates; therefore, permeability and friability are better in aggregate soil conditions.

Thus, soil solutions composed of high concentrations of solutes or with a predominance of calcium and magnesium salts provide soils with good physical properties. Conversely, low salt concentrations and relatively high proportions of sodium salts drastically affect permeability and hydraulic conductivity, while a high pH value (>8) also adversely affects these characteristics, because it increases the amount of negative charges in the clay and organic matter and also the repulsive forces between them (RHOADES; KANDIAH; MASHALI, 1992).

During infiltration, the soil solution in the top layer is essentially that of the infiltration water and the percentage of exchangeable sodium is similar to that present in the soil. Because all the water that enters the soil passes through the surface of the soil, and because is more subject to loss by aggregation, the properties of the topsoil greatly control the speed of water infiltration; These observations, taken together with knowledge of the effects of the processes listed above, explain why problems of permeability and hydraulic conductivity should be evaluated in terms of both the salinity of the infiltrated water and the percentage of exchangeable sodium (or its value equivalent to RAS) and the pH of the topsoil (RHOADES; KANDIAH; MASHALI, 1992).

Decreases in the infiltration rate (VI) of the soil generally occur during the irrigation season, due to the gradual deterioration of the structure and formation of a surface seal (horizontal layering of soil particles) as a result of successive irrigations (sedimentation, wetting and drying events). VI is even more sensitive to exchangeable sodium, electrolyte concentration and pH than hydraulic conductivity. This is due to the significant vulnerability of the topsoil to mechanical forces, which increase the dispersion of clays, the disintegration of aggregates and the movement of disaggregated clay in the soil near the surface, as well as the low electrolyte concentration that generally exists there, particularly under rainy conditions. Crusts, due to deposition, often occur in irrigated furrows where soil particles, suspended in water, are deposited when the water flow is low or when the water infiltrates (RHOADES; KANDIAH; MASHALI, 1992).

In soils from arid regions, the use of cation exchange equations to show the relationship between cations in solution and adsorbed cations involves its own difficulties, due to the mixture of different types of clay in the soil, as well as the presence of four species of cations. In the 1930s and 1940s, some authors (GAPON; MATTSON; WIKLANDER; DAVIS; SHOFIELD) showed that the function between the ratio of adsorbed monovalent and bivalent cations and the ratio between the molar concentration of the monovalent cation and the square root of the molar concentration of the bivalent cation is linear (RHOADES, KANDIAH; MASHALI, 1992).

3.4 **EFFECTS OF SALTS ON PLANTS**

The process of salinity in the root zone generally has a deleterious effect on plant growth, manifested as an equivalent reduction in transpiration and growth rates (including cell expansion, synthesis of metabolites and structural components). This effect is primarily related to electrolyte concentration and is independent of the specific composition of the solute. The hypothesis that seems to best fit the observations is that excessive salinity reduces plant growth by causing an increase in energy expenditure to absorb water from the soil and carry out the biochemical processes needed to survive under stressful conditions. This energy is diverted from the processes that lead to growth and production (RHOADES; KANDIAH; MASHALI, 1992).

Plants can be classified from tolerant to sensitive to salinity. In some cases, they can be called halophytes, as they are able to absorb salts by establishing an osmotic balance so that there is no interference with the enzymatic, metabolic and hydration mechanism (LAUCHI; EPSTEIN, 1984). Plants that are sensitive to salinity generally tend to exclude salts when absorbing soil solution, thus being limited in terms of osmotic adjustment. The consequence of this is osmotic stress, causing the cells to become less turgid (BARROS, 2002).

Growth suppression typically begins at some threshold value of salinity, which varies with the tolerance of the crop and external environmental factors that influence the plant's need for water, in particular the evaporative demand of the atmosphere (temperature, relative humidity, wind speed, etc.) and the potential supply of the root zone. This suppression increases as salinity increases, until the plants die. The tolerance of various crops to salinity is conventionally expressed (MAAS; HOFFMAN, 1977) in terms of relative yield (Yr), threshold salinity value (a) and point decreases in production (b) per unit increase in salinity above the threshold salinity where soil salinity is expressed in terms of ECe, in dS m^{-1}, as follows: Yr=100- b(CEe-a), where Yr is the percentage of yield expected under saline conditions, relative to that obtained under non-saline conditions, kept comparable to the other conditions. This use of ECe to express the effect of salinity on production implies the fact that plants respond primarily to the osmotic potential of the soil solution (RHOADES; KANDIAH; MASHALI, 1992).

It is important to consider that such salt tolerance data cannot accurately provide quantitative yield losses as a result of salinity for all situations, since the response of plants to salinity varies with other growing conditions, such as climatic and soil conditions, agricultural and irrigation management, crop variety, growth stage, etc.Climate is the factor that most affects salt tolerance; many crops can tolerate great salt stress if the weather is cool and wet, rather than hot and dry. Production is more reduced by the effect of salinity when the humidity of the atmosphere is low (RHOADES, KANDIAH; MASHALI, 1992).

In general, plants are relatively tolerant to salinity during germination, but they become more sensitive during emergence and the initial growth stage. Tables 1 and 2 show examples of how some crops tolerate salinity.

Table 1. Relative tolerance of different crops to salinity at emergence and during growth and maturation (Maas, 1986).

Cultures		Electrical Conductivity of the Soil Saturation Extract	
Common Name	Scientific name	50 % yields dSm^{-1}	50 % Emergence[1] dSm^{-1}
Cowpeas	*Vigna unguiculata*	9,1	16
Corn	*Zea mays*	5,9	21-24
Sorghum	*Sorghum bicolor*	15	13

[1]Percentage emergence of the salt treatments determined when the non-salt treatments had reached maximum emergence.

Table 2. Tolerance of herbaceous crops to salinity (Maas, 1986)

Culture		Electrical Conductivity of Soil Saturation Extract		Class
Common Name	Scientific name	Threshold[3] dSm-1	Slope % dSm^{-1}	
Melon	*Cucumis melo*	1,0	6,4	MS*
Cowpeas	*Vigna unguiculata*	4,9	12,0	MT*
Corn	*Zea mays*	1,7	12,0	MS
Sorghum	*Sorghum bicolor*	6,8	16,0	MT
Pepper	*Capsicum annuum*	1,5	14,1	MS

*MT= moderately tolerant, MS= moderately sensitive. Data referring to crops irrigated by surface methods (furrow and flood) and conventional irrigation management.

Significant differences in salt tolerance occur between varieties of some species, although this issue is confusing because of the different climatic or nutritional conditions under which the crops have been tested and the possibility of better varietal adaptation to these conditions. Salt tolerance also depends on the type, method and frequency of irrigation. As the soil dries out, matric and osmotic stress occurs, limiting water absorption. Plants respond to salinity in that part of the root zone where water absorption is greatest. Thus, tolerance is related to salinity weighted over a certain period and measured where the roots absorb the most water (RHOADES, KANDIAH;

MASHALI, 1992).

Visible symptoms can also occur suddenly after irrigation when the weather is hot and dry. Thus, an increase in the frequency of sprinkler irrigation, together with an increase in temperature and evaporation, leads to an increase in the concentration of salts in the leaves and leaf damage.

In herbaceous crops, the main effect of salinity is to stunt growth, because certain saline constituents are specifically toxic to some crops. Boron is one of them, and when present in the soil solution in concentrations of a few mg L^{-1}, it is highly toxic to susceptible crops.

For some crops, especially trees and perennials, sodium and chloride can accumulate in the tissues over time until they reach toxic levels, with consequent leaf burn. Generally, these plants are also sensitive to salinity, making it difficult to separate the two effects.

Increases in salinity can cause losses in the production and quality of fruit, vegetables and grains, with changes in size, color, appearance and chemical composition.

3.4.1 Effect of salts on the melon crop

The melon crop is classified by Ayers and Westcot (1991) as moderately sensitive to salinity, showing low yields at certain times of the year, when there is a greater accumulation of salts in the water table and in reservoirs. Feigin (1990), studying the responses of Ganglia melon plants under saline (9.0 dSm^{-1}) and non-saline (1.5 $dSm^{-1)}$ conditions in a nutrient solution, observed a significant reduction in dry matter weight. Chartozoulakis (1992) reported that values above 5 dSm^{-1} normally cause a reduction in germination speed. For Sivritepe *et al.* (2003), high concentrations of salts in soil and water cause morphological, structural and metabolic changes in melon seedlings, inhibiting their growth and development, reducing the percentage of dry mass in plants, fruit size and yield (MENDLINGER; FOSSEN, 1993). Meiri *et al.* (1981) observed that increased salinity reduced fruit size and accelerated ripening. According to these authors, melons vary greatly in terms of their tolerance to salinity, both between cultivars and in terms of environmental and management conditions. According to Pizarro (1990), on average the melon has a threshold salinity of 2.2 dSm^{-1} and a decline in production per unit increase in the electrical conductivity of the soil saturation extract of 7.25 dSm-1. It can therefore be seen that, in general, salinization negatively affects germination, plant stand, the vegetative development of crops, productivity and, in the most serious cases, causes seedling death (SILVA and PRUSKI, 1997).

Silva *et al.* (2005), studying the fruit yield of the melon tree under different levels of irrigation water salinity, showed that the total yield was reduced as the level of water salinity increased, which is common for almost all crops.

As the melon tree has a moderate tolerance to salinity, irrigation with high salinity water (up to 2.5 dS $m^{-1)}$, provided it is well managed and in soil with good drainage, can produce sweeter fruit with a high soluble solids content without producing significant reductions in productivity (CUENCA, 1989). However, during the initial growth phase, the plant is more sensitive to salinity and there is evidence that high soil salinity during the flowering period can reduce the number of fruits set (CUENCA, 1989).

Hά the need to develop melon cultivars with salt tolerance, due to the increase in soil salinity (SIVITREPE *et al.*, 2003). In semi-arid conditions, where there is a constant increase in soil salinity due to the water used for irrigation, there is a need for melon cultivars that can produce with economic viability, meeting the growing demand for this fruit.

3.4.2 Effect of salinity on the caupi bean crop

The cowpea (*Vigna unguiculata* L. (Walp.)) is a species well adapted to the conditions of the semi-arid region where adverse conditions of drought, salinity, high temperatures and high insolation often prevail. These stress factors interact with each other, determining complex responses in the plant (RIBEIRO, 2002).

This legume is one of the main components of the northeastern population's diet and its grains have a high nutritional value due to their high protein content (LIMA *et al.*

2007).

According to data obtained by the FAO, the string bean tolerates irrigation with saline water with an electrical conductivity of up to 3.3 dS m^{-1} (AYERS and WESTCOT, 1999), and is therefore considered to be a species that is moderately tolerant to salinity. However, Dantas *et al.* (2002) state that the degree of tolerance of caupi to saline stress varies between genotypes.

Working with the Epace 10 cultivar of cowpea, Assis Júnior *et al.* (2007) found that the reduction in productivity caused by salinity was partly due to reductions in vegetative growth and net carbon assimilation, during the flowering and fruiting phases, associated with osmotic effects and the accumulation of potentially toxic ions (Na^+ and Cl^-) in leaf tissues.

Testing the salinity tolerance of the *Quarentinha* cultivar of cowpea, Lima (2007) found that the dry matter of the aerial part of the plant decreased linearly as the salinity of the irrigation water increased, reaching a reduction of 66.94% at the highest saline level (5.0 dS m-1) in relation to the control (0.5 dS m-1).

In evaluating the growth of three vigna bean cultivars subjected to saline stresses of 2 and 10 dS m-1, Patel (2010) found that there was a reduction in the dry biomass of the aerial part of the plant as salinity levels increased, in the order of 33, 39 and 43% for the cultivars Akshay-102, Gomtivu-89 and Pusa Falguni, respectively.

Some studies have shown that the use of biofertilizers in saline environments can partially mitigate the effects of salinity on plant growth (BEZERRA *et al*, 2010; CAVALCANTE, 2010; OULD AHMED and MORITANI, 2010). The importance of using liquid biofertilizers in the form of simple or enriched microbial fermentates lies in the quantity of elements, the diversity of mineral nutrients and the availability of nutrients through biological activity (ALVES, 2009).

3.4.3 Effect of salinity on maize crops

According to Ayers (1991), sweet corn and forage corn are classified as moderately sensitive to irrigation water salinity, since for this crop to reach 100% production, the electrical conductivity (EC) in the soil must be 1.7 dS m^{-1} and EC in the water 1.1 dS m^{-1}; to obtain 0% production, the EC in the soil must reach ά 10.0 dS m^{-1} and in the water 6.2 dS $m^{(-1)}$.

When evaluating the initial development of popcorn maize 'Zélia hybrid' irrigated with different levels of irrigation water salinity, Oliveira (2009) found that an increase in water salinity inhibited the diameter of the stem and the number of leaves the plants produced. In this same species, Garcia (2007), Blanco (2008) and Gondim (2011) observed that salt stress affected the dry matter content of the aerial parts and roots; while Sousa, (2010) in the hybrid AG 1051 obtained reductions in the accumulation of biomass of the vegetative parts (leaf + stalk).

Garcia *et al*, (2007) found that the increase in soil salinity resulting from irrigation with saline water increased sodium levels, the Na^+/Ca^{2+}, $Na^+/Mg^{(2+)}$ratios, Na^+/K^+ and reduced the levels of calcium, magnesium and potassium, thus reflecting the nutritional imbalance caused by the nutritional stress resulting from progressive saline stress.

3.4.4 Effect of salinity on sorghum crops

According to Taylor (1975) and Fançois (1984), sorghum has a recognized ability to adapt to soils with salinity problems. According to Ayres and Westcot (1999), sorghum is moderately tolerant to salt stress and could be an alternative for crops that use saline resources, which is justified by its bromatological characteristics which, like maize, This is justified by its bromatological characteristics which, like maize, enable adequate fermentation and consequent storage in the form of silage, by the high levels of soluble carbohydrates and even crude protein in some varieties, and by its agronomic characteristics which, among others, include moderate resistance to drought (TABOSA, 2002) and high biomass productivity (OLIVEIRA, 1983).

Selections of forage sorghum genetic materials for salinity tolerance have been made in terms of their adaptability to saline conditions in the field (FERNANDES, 994), the growth of plants in nutrient solution in the greenhouse (SILVA, 2003) and the leaf growth of plants grown in nutrient solution under controlled conditions (LACERDA, 2003).

In an experiment evaluating forage sorghum cultivars in soils affected by salts in the state of Rio Grande do Norte, it was found that the cultivars behaved similarly under low salinity conditions in the municipalities of Mossoró and Ipanguaçu. However, when subjected to high salinity conditions, such as in Apodi, there was a difference between the cultivars in terms of dry mass accumulation, with IPA 1218 standing out. The average annual production of green mass has shown a downward trend over the years of cultivation (FERNANDES, 1994).

3.4.5 . Effect of salinity on pepper crops

The pepper crop is characterized by being moderately sensitive to salinity (MAAS, 1986). Under conditions of saline stress, the colour of pepper leaves (*Capsicum annuum* L.) can become dark bluish-green, with greater thickness and waxyness; in the roots there can be a decrease in elongation and suberization, which affects the absorption of water and nutrients, with a consequent reduction in productivity (CRUCIANI, 1987).

Leonardo (2007) observed that the salinity levels of irrigation water correlated inversely with the setting index and average weight of pepper fruit. For the treatment with the highest salinity level (6 dS m^{-1}), the fruit showed a reduction of up to 58% in average weight and 55% in the number of fruits per plant, compared to the control treatment, which showed that the EC in the soil solution drastically interfered with the production parameters evaluated.

3.5 THE INFLUENCE OF SALINITY ON PLANT MINERAL NUTRITION

The high concentration of salts is a stress factor for plants, since it has an osmotic activity that retains water, in addition to the action of ions on protoplasm and the impact on plant mineral nutrition (MUNNS, 2002). Salinity affects the mineral nutrition of crops by reducing the activity of ions in solution and altering the processes of absorption, transport, assimilation and distribution of nutrients in the plant. The interaction between salinity and mineral nutrition becomes more complex due to the differences in the concentration and ionic composition of the saline media (water and soil) to which plants are subjected, and the different responses of plants, both in relation to salinity and in relation to the efficiency of mineral acquisition from the soil (LACERDA, 2005). It is possible to suggest, however, that the predominance of certain mineral elements in the root environment, as is common in saline environments, directly or indirectly affects the acquisition of essential minerals and can lead to a reduction in growth and alterations in the quality of the plant product (GRATTAN; GRIEVE, 1999; LACERDA, 2005).

According to Bataglia (1996), during the initial phase of the vegetative period, the rate of nutrient absorption is high and, consequently, the concentrations in the tissue are also high. As the plant grows, there is dilution, decreasing the concentration of nutrients in the tissue. Thus, physiologically young tissues generally have higher concentrations. In general, the levels of N, P and K decrease with the age of the plant, while the levels of Ca, Mg, Mn and B often increase; however, in this study, Mg behaved in a similar way to N, P and K, i.e. it was diluted from the age of six months.

There is still a lack of studies in the literature showing the amount of elements extracted by crops at different stages of development under saline stress. The ionic interactions that affect the availability, absorption and transport of nutrients are highly complex, even in the absence of salinity and other stresses. Salinity adds a new level of complexity to crop mineral nutrition, affecting the activity of ions in solution and the processes of absorption, transport, assimilation and distribution (NEVES , 2009).

3.6 PRACTICES TO REDUCE THE EFFECT OF SALINITY

In many cases, values of electrical conductivity of the saturation extract above 0.7 dS m^{-1} cause severe damage to a large number of plants subjected to continuous irrigation. Thus, Ayers and Westcot (1999) highlight some procedures that can be used to reduce salinity problems, such as: salt leaching, choice of irrigation method, underground drainage, fertilization and agronomic management.

The leaching rate that controls the accumulation of salts in the root zone of irrigated crops depends on the saline concentration of the water, the method of application, rainfall and the physical and chemical characteristics of the soil (PIZARRO, 1985). The level of salts in the root zone must be below the level harmful to plants, so the amount of salts removed must be equal to or greater than that applied via irrigation, to avoid accumulation and concentration at dangerous levels.

The choice of irrigation method affects both water use efficiency and soil salinity. In the case of furrow irrigation, salts accumulate at the top of the furrow, so sowing on the furrow slope favors the formation of a good plant stand. On the other hand, drip irrigation maintains a high level of humidity in the plant's root zone, which reduces the effect of salts on germination and seedling establishment. When saline water is used, drip, furrow, flood and micro-sprinkler irrigation methods give better results than sprinkler irrigation, since this method can lead to salt accumulation and toxicity in the leaves of some species (SHARMA; MINHAS, 2005). In this respect, drip irrigation is the best method, since the water potential is kept close to zero in the root environment, which reduces the osmotic effects of the salts on the plant, in which case there may be a long-term accumulation of salts in the soil.

The lack of adequate surface drainage and the presence of a shallow water table aggravate salinity problems and complicate their control. Often, the presence of a shallow water table is associated with the presence of low-permeability soil, such as a layer of heavy clay or the presence of bedrock. Excess water often comes from excessive irrigation, filtration from higher ground and/or water leaking from canals. If you irrigate soil with poor or no drainage and in regions with low rainfall, it becomes saline in a fairly short period of time, because the plants basically remove the water while most of the salts are retained. The most effective way of controlling salinity resulting from shallow water tables is through a drainage system that allows the water level to be lowered and kept stable at this depth (LIMA JÙNIOR; SILVA, 2010).

Natural and chemical fertilizers and soil improvers contain high concentrations of soluble salts. When placed near seeds or plants, these products cause or intensify salinity and/or toxicity problems. Plants in the early stages of development are sensitive to salinity and require little fertilization; therefore, a small amount of fertilizer can be applied before or at the time of sowing and the rest in one or more applications between emergence and the vigorous growth phase. When fertilizing, consider using less saline sources. High levels of nutrients in saline environments and maintenance of absorption can lead to the accumulation of nutrients in the tissues, resulting in a lack of adjustment between acquisition and assimilation, leading to toxicity and intensifying the deleterious effects caused by salinity. In addition to conventional fertilization techniques with mineral fertilizers, the use of alternative forms that favour the acquisition of nutrients by the plant, such as the application of liquid biofertilizers, organic matter and the inoculation of plant roots with mycorrhizal fungi has shown an increase in the development and productivity of some crops (CAVALCANTE, 2010; SILVA, 2011).

With regard to agricultural management, some practices can be used to minimize the effect of salinity. Plowing or subsoiling can redistribute salts in the soil profile, so that layers of poorly permeable soil that prevent the movement of water through the root zone can be distributed; however, this practice has a temporary effect. Applying manure and crop residues to the soil promotes soil structure and improves water infiltration. Residues can be left as mulch on the surface or incorporated into the soil. The use of mulch for various crops is an important practice, controlling the balance of radiation and providing greater vegetative and productive development, reducing damage to the skin of the fruit, controlling a large part of the weeds and also increasing tolerance to the use of water with high salinity (NASCIMENTO *et al*, 2000). The mulch, by eliminating evaporation from the soil surface, allows the plant to absorb water from this layer, a region with a lower saline concentration, and this improves the absorption of fertilizers applied in fertigation, as this is where it arrives first, together with the irrigation water. The effects are also temporary, requiring repetition periodically.

The accumulation of salts during the irrigation of crops in the dry season can be totally or partially reversed during the rainy season, and this process depends on the intensity of rainfall and the physical characteristics of the soil (SHARMA; RAO, 1998). It is important to note that in semi-arid conditions, the total rainfall may not be sufficient to wash salts out of the soil (MEIRELES *et al.*, 2003). In addition, the use of chemical conditioners is recommended as a preventative measure whenever sodium-rich water is used, in order to increase the efficiency of the sodium leaching process and reduce its impact on the soil's physical properties.

An alternative that can reduce the concentration of salts and increase the availability of water for irrigation is the mixing of water of different qualities (MALASH *et al.*, 2005). However, mixing water is only feasible when more than one water source is available, and this is possible in many countries.

practical situations, particularly when surface and groundwater sources are available. Alternatively, you can opt for the cyclical use of water sources with different salt concentrations (FLOWERS *et al.*, 2005). This strategy reduces the impact of salinity on the soil and the plant, as well as increasing the efficiency of using good quality water.

4. MATERIAL AND METHODS

4.1 LOCATION AND DESCRIPTION OF THE STUDY AREA

Two experiments were carried out over a period of 12 months in a greenhouse at the Department of Environmental and Technological Sciences at UFERSA, Mossoró Campus - RN, with the following geographical coordinates: south latitude 5° 11', west longitude 37° 20' and altitude 18 m. The dominant climate in the region, according to the Koppen classification, is the BSw h' type, i.e. hot and dry. The region's dominant climate, according to the Koppen classification, is *BSw* h', i.e. hot and dry, characterizing it as a semi-arid tropical climate, with a very irregular rainy season, delayed from summer to autumn and concentrated in the first few months of the year. According to Carmo Filho and Oliveira (1989), the characteristics of the location are: average temperature of 27.4 °C, average annual rainfall of 673.9 mm and relative humidity of 68.9%. The greenhouse where the experiments were carried out has the following characteristics: a metal structure with a ceiling height of 3.5 meters, seven meters wide and 18 meters long with an arched roof and transparent plastic treated against the action of ultraviolet rays, closed on the sides with a 50% sombrite screen (Figure 1).

Figure 1. Vegetation house where the experiments were carried out. Mossoró, 2010

Materials from five soils that occur most frequently in the Mossoró-Assu agropolis were used: Quartzarenic Neosol (RQ) - Tibau-RN (Fazenda Famosa), Argissolo Vermelho Amarelo Eutrófico latossólico (PVAe) - Mossoró- RN (Fazenda Alaginha), Cambissolo Hàpico (CXve) - Baraùnas -RN (Fazenda pH), Neossolo Flùvico (RY) - Assu-RN (Sitio Sao Mateus) and Vertissolo (VX) - Mossoró-RN (Assentamento Paulo Freire), which were collected by digging trenches in the 0-20 and 20-40 cm depth layers and physically and chemically characterized at the UFERSA soil laboratory, according to the EMBRAPA methodology (1999), the results of which are shown in Table 3. After being dried and sieved, the soil materials were distributed in PVC columns with a diameter of 20 cm and a height of 50 cm, with a capacity for 12 liters of soil material.

Table 3. Chemical and physico-hydric characterization of the soils used in Experiments I and II. Mossoró-RN, 2010

		Soils									
		RQ		PVAe		CXve		RY		VX	
					Profu ıdity d	and collection					
Features	Und.	0-20	20-40	0-20 :	20-40	0-20	20-40	0-20	20-40	0-20	20-40
Total N	g/kg	0,56	0,14	0,14	0,42	1,05	0,63	0,42	0,49	0,42	0,07
MO	g/kg	12,0	3,00	4,00	6,60	15,0	7,60	7,00	5,60	7,40	4,20
pH		5,25	5,30	7,10	6,85	7,56	7,11	7,33	7,16	8,63	9,10
P	$mgdm^{-3}$	6,40	1,60	3,00	6,70	6,50	2,30	76,1	88,0	3,10	0,80
K^+	$mg\ dm^{-3}$	53,2	18,3	39,2	66,1	335,2	270,4	146,9	124,9	41,12	27,2
In^+	$mg\ dm^{-3}$	16,6	10,7	11,6	14,6	15,6	15,6	96,2	88,4	83,4	30,27
Ca^{2+}	$cmolcdm^{-3}$	2,58	0,73	1,29	1,81	7,99	7,90	13,2	9,67	36,1	39,72
Mg^{2-}	$cmolc\ dm^{-3}$	0,52	0,44	0,35	0,40	1,99	0,00	0,71	4,18	10,54	6,52
Al^{3+}	$cmolc\ dm^{-3}$	0,00	0,00	0,00	0,00	0,00	0,00	0,00	0,00	0,00	0,00
Cu	$mg\ dm^{-3}$	0,28	0,21	0,27	0,21	1,61	1,54	0,94	1,02	0,18	0,18
Fe	$mg\ dm^{-3}$	25,9	20,8	7,70	8,20	8,80	6,40	27,9	21,5	3,80	1,80
Mn	$mg\ dm^{-3}$	2,60	1,90	4,10	10,30	12,2	13,7	18,2	19,3	1,40	1,60
Zn	$mg\ dm^{-3}$	1,10	0,75	2,05	1,47	2,56	2,93	0,75	1,34	0,83	0,85
H+Al	$cmolc\ dm^{-3}$	2,81	1,24	0,00	0,58	0,00	0,00	0,00	0,00	0,00	0,00
SB	$cmolc\ dm^{-3}$	3,31	1,26	1,79	2,44	10,91	8,66	14,7	14,6	47,1	47,6
T	$cmolc\ dm^{-3}$	3,31	1,26	1,79	2,44	10,91	8,66	14,7	14,6	47,1	47,6
CTC	$cmolcdm^{-3}$	6,11	2,50	1,79	3,02	10,91	8,66	14,7	15,6	47,1	47,6
V	%	54	51	100	81	100	100	100	100	100	100

M	%	0	0	0	0	0	0	0	0	0	0
PST	%	1	2	3	2	1	1	3	3	1	3
Sand	g kg^{-1}	930	930	930	940	510	350	490	500	420	370
Silt	g kg^{-1}	40	10	10	20	150	160	370	350	250	340
Clay	g kg^{-1}	30	60	60	40	340	490	140	160	330	300
Depth (cm)			0-40		0-40	0-40		0-40		0-40	
Soil density	g cm^{-3}		1,31		1,61	1,55		1,45		1,32	
Part density	g cm^{-3}		2,63		2,66	2,58		2,47		2,47	
Sat humidity	%		34		37	53		41,30		46,56	
Porosity	%		50		39,5	40		29,5		37,8	
Water available	%		0,4		0,4	6,2		30,2		25	

Analyses carried out at UFERSA's Soil Fertility and Fertilization and Salinity laboratories.

The samples were subjected to tensions of 10, 20, 30, 40 and 60 cm.c.a. on a sand porous plate tension table, to a tension of 100 cm.c.a. on a clay porous plate tension table and to tensions of 800, 2000 and 15000 cm.c.a. in a pressure chamber with a porous plate. After reaching equilibrium for each stress, the samples were weighed to determine their wet mass, after which they were resaturated at their base just to re-establish hydraulic contact so that they could be subjected to the next stress. After all the tensions had been applied, the samples were dried in ovens at 105 °C for 48 hours to determine their mass of solids. This information was used to calculate the volumetric humidity and density of the soil. The data was adjusted to Van Genuchten's model (1980) using the computer program Soil Water Retention Curve SWRC version 3.0 beta (DOURADO NETO *et al.*, 2001), applying Mualem's theory (1976).

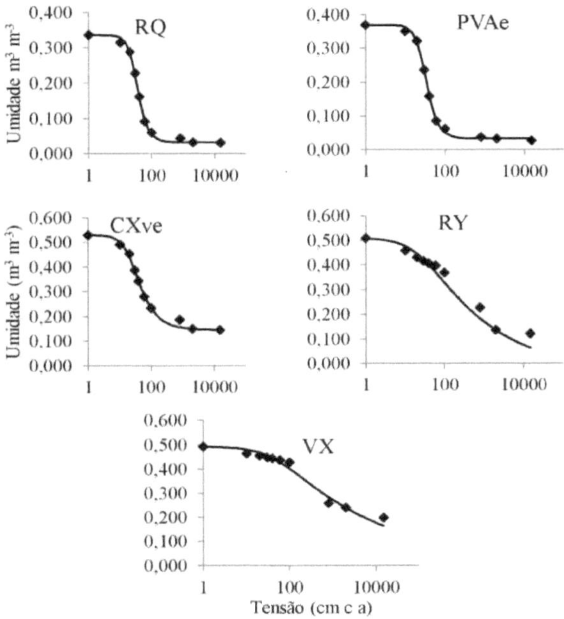

Figure 2. Water retention curve in the soils used in the experiments, averages of the 020 and 20-40 cm layers. Mossoró-RN, 2010.

The waters used during the experiments came from two aquifers: the Açu Sandstone, with a depth of approximately 1000 m, characterized by low salinity and used for local urban supply; and the Jandaira Calcareous Aquifer, with a depth of approximately 150 m, with saline water and the main predominant cations being Na^+ and Ca^{++} and the anions HCO3" and Cl^-. The chemical characteristics of the waters used in the experiments are shown in Table 4.

Table 4. Chemical characteristics of the waters used during the experiments. Mossoró-RN 2010

Chemical characteristics	Units	Ufersa water supply	Water from the Zootechnics shallow well
ECa	dS m^{-1}	0,59	4,33
Ph	-	7,50	6,90
K	mmolcL^{-1}	0,53	0,15
In	mmolc L^{-1}	2,79	19,00
Ca	mmolc L^{-1}	1,80	15,20
Mg	mmolc L^{-1}	0,50	2,80

CO_3	mmolc L^{-1}	0,35	0,20
HCO_3	mmolc L^{-1}	4,10	4,80
Cl	mmolc L^{-1}	1,60	25,20
∑CAtions	mmolc L^{-1}	5,62	37,15
£Âni *o* ns	mmolcL^{-1}	6,05	30,20
RAS	$(mmolL^{-1})^{0.5}$	0,74	6,33
Classif.*		C2S1	C4S1

* Richards (1954)

4.2 TREATMENTS AND EXPERIMENTAL DESIGN

The treatments used in experiment I were a combination of two factors: five types of soil (RQ, PVAe, CXve, RY and VX) and six levels of irrigation water salinity (A_1= 0.59; A_2= 1.00; A_3= 2.00; A_4= 3.00; A_5= 4.0 and A_6= 5.0 dS $m^{-1)}$.The statistical design used was entirely randomized in a 5x6 factorial scheme (soils x salinity) with 3 replications, totaling 90 experimental units. The saline concentrations were obtained by mixing water of higher salinity (EC = 4.33 dS m^{-1}) from a shallow well with water of lower salinity (EC = 0.59 dS m^{-1}) from a deep well and to complete the salinity up to the highest level (5.0 dS m^{-1}) NaCl salt was added to increase the ECa to 5.0 dS $m^{(-1)}$. Each plot consisted of two plants per column of soil material, distributed on the benches, spaced 0.5 m apart and 1.0 m between the benches.

In experiment II, the treatments were made up of a combination of five crops: Melon, Caupi Beans, Maize, Sorghum and Pepper; two types of soil material: PVAe and CXve and three levels of irrigation water salinity (A_1=0.59; A_2=2.75 and A_3= 5.0 dS m^{-1}). A completely randomized design was used in a 3x2 factorial scheme with three replications, totaling 90 experimental units. The same procedures were used to obtain the salinity levels as in the first experiment.

4.3 DESCRIPTION OF THE CULTIVARS STUDIED

In experiment I, the Mandacaru hybrid melon was used. This hybrid was chosen because it had the largest growing area in the region, as well as being well adapted to the region's climatic conditions and having desirable agronomic characteristics for the domestic and foreign markets. Experiment II used melon (cv. Néctar), beans (Epace

10), corn (Impacto), sorghum (BR 304) and pepper (Atlantis). These crops were chosen because they are the most widely used by producers and have an expanding area of cultivation in the region, as well as due to the lack of technical information, especially regarding salinity tolerance, in addition to having good adaptation to the region's climatic conditions and desirable agronomic characteristics, as described in the topics below.

4.3.1 Melao

The Yellow melon cultivar (Figure 3), which belongs to the odorless group, has high productivity and vigorous plants. The fruits of the Mandacaru melon variety are characterized by: yellow skin, creamy white flesh, rounded fruit shape, medium to large size with an average weight ranging from 1.5 to 2.3 kg. It has a cycle time that is considered average for the climate conditions of the Northeast (CLOUSE, 2009).

Figure 3 - Mandacaru cultivar. Mossoró-RN, 2010

The Gàlia type used in Experiment II includes reticulated aromatic melons of Israeli origin. The fruit in this group is characterized by its round shape, green skin that changes to yellow when the fruit ripens, white or greenish-white flesh, little cross-linking and an average weight of between 0.7 and 1.3 kg. According to Clouse (2009), the Néctar variety has the following characteristics: rounded fruit weighing 0.8 to 1.2 kg, lacy skin, green and very sweet flesh, and excellent post-harvest preservation of the fruit (Figure 4). The plants are medium vigor, with high production potential.

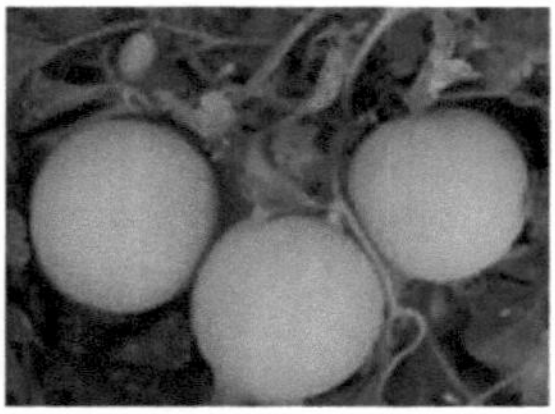

Figure 4 - Nectar cultivar. Mossoró-RN, 2010

4.3.2 Cowpeas

The EPACE-10 cowpea cultivar was developed at the Cariri Research Department to expand the genetic possibilities linked to production potential, resistance to viruses, tolerance to drought and commercial quality of the grain and pod. Its characteristics are indeterminate growth habit, semi-branching size, 65-75 day cycle, violet flower color, pod length 21 cm, 15 grains per pod, brown seed, yield 1000 kg ha^{-1}.

Figura 5. Bean cv. Epace 10. Mossoró-RN, 2010

4.3.3 Maize

The hybrid maize cultivar Impacto is characterized by exceptional leaf health against the main diseases, especially *Cercospora*, providing great performance in the summer in central Brazil, reducing the population of nematodes. It has a hard, yellow-orange grain with an excellent reproductive ceiling and stability. It is used in winter and summer plantings. With a population of 50 to 65 thousand plants ha^{-1} in the summer and 50 to 55 plants ha^{-1} in the winter.

Figura 6. Maize cultivar Impacto. Mossoró-RN, 2010

4.3.4 Sorghum

Cultivar BR 304 is a simple hybrid, the result of crossing BR 001 A with BR 012 R. In succession plantings, BR 304 is productive and tolerant to dry rot of the stalk, caused by the pathogen *Macrophomina phaseolina*, under conditions of water deficiency. This characteristic makes BR 304 highly recommended for planting in succession to early summer crops and for late plantings. It is recommended for the Northeast, Center-West, Southeast and South regions, especially for times when rainfall is less available.

Figura 7. Sorghum cultivar BR 304. Mossoró-RN, 2010

4.3.5 Pepper

The Atlantis pepper cultivar is a simple hybrid, with green or red fruits weighing 240 to 260 grams, very thick walls of 6 to 8 mm and excellent post-harvest. The plant is vigorous, with good leaf cover which guarantees good fruit protection. It has thick stems and short internodes. It is characterized by being a variety that can be grown in field or greenhouse conditions with high production potential.

Figure 8. Atlantis pepper. Mossoró-RN, 2010

4.4 CONDUCTING THE EXPERIMENTS

After distributing the soil material in the columns, they were fertilized with 150 ppm of P2O5 and watered until they reached field capacity. This was followed by direct sowing, according to the crops planned for each experiment.

The volumes of irrigation water were determined based on estimates of crop evapotranspiration and adjusted by monitoring soil moisture using tensiometers which were distributed radially at an average distance of 0.15 m from the plant and at depths of 0.15 and 0.3 m. The tensiometers were used as soil moisture extractors. The tensiometers also served as soil solute extractors. FDR probes were also installed to simultaneously measure salinity and volumetric soil moisture according to Equation 1 of Topp *et al.* (1980), which was adjusted in the laboratory.

Equation 1 was used to calculate the potential of the water in the soil, determined using a tensimeter.

$$\square_m = (1020\ L_T) + (20 + Z) \qquad (1)$$

where:

$\square_m$= soil water potential (kPa);

L_T = tensimeter reading (bar);

Z = depth from the soil surface to the center of the ceramic capsule (cm); 20 = height of the water column, from the soil surface to the water surface inside the tensiometers.

The matric potential of the soil water (Ψm), at a depth of 0.15 m, served as a critical value to indicate when to irrigate. Both the tensiometers located at a depth of 0.15 m and at the other depths were used to quantify the water stored in the soil.

Fertigation was carried out weekly, with the nutrients being distributed throughout the initial development of the crops according to their nutritional needs, following the absorption pattern of each crop, using the following fertilizers: potassium chloride, potassium sulphate, urea, calcium nitrate and magnesium sulphate.

Pest and disease control has always been preventative, with the aim of avoiding the presence of the main pests and diseases common in the region, through the use of specific insecticides and fungicides for each crop.

The melon plants and other crops were grown until flowering, when the aerial part was cut off close to the ground and taken to the Irrigation and Salinity laboratory for growth, dry matter and nutritional analyses.

In experiments I and II, two leaching rates were applied to wash salts out of the soil profile, applying a leaching fraction of 10%. The climatic data for the period during which the experiments were conducted is described in Table 5.

Table 5. Data from UFERSA's automatic weather station for the period during which the experiment was conducted. Mossoró-RN, 2010

Week	T. med	T. max	T. min	RH (%)	V2 ($m\ s^{-1}$)	RG ($MJ\ m^{-2}$)	ETo ($mm\ day^{-1}$)
	------	----- °C------	------				
1	27,95	35,52	20,38	54,00	5,42	20,58	7,77
2	27,29	34,79	19,80	51,04	6,19	20,16	8,35
3	28,45	35,84	21,07	50,09	5,59	21,97	8,67
4	28,50	35,81	21,18	55,40	5,80	20,50	8,19
5	28,65	36,00	21,30	51,45	5,89	20,54	8,65
6	28,71	35,93	21,49	51,70	6,38	23,54	9,26
7	28,76	33,62	23,90	66,17	6,46	21,00	7,51
8	29,98	36,41	23,54	57,60	6,93	24,45	9,41
9	29,88	36,38	23,37	54,65	7,23	23,26	9,63
10	29,29	35,25	23,34	60,45	6,82	22,99	8,64
11	29,41	34,92	23,90	60,44	7,24	23,88	8,89
12	29,62	35,78	23,45	57,49	7,27	21,31	8,94
13	28,70	33,29	24,11	66,95	6,21	17,36	6,76
14	29,20	34,00	24,39	67,94	5,87	17,95	6,79
15	29,17	35,19	23,14	66,50	5,98	19,36	7,22
16	28,63	34,42	22,85	63,57	6,74	23,26	8,16

Source: UFERSA Automatic Weather Station. Mossoró, RN, 2010

T. Average, T. Maximum and T. Minimum→ Average, maximum and minimum temperature; UR→ Average daily relative humidity; v2→ Average daily wind speed at a height of 2 meters; RG → Average global radiation,

ETo - FAO → Reference evapotranspiration estimated by the Panmam Monteih equation - FAO.

Table 6. Average irrigation volume values for fortnightly periods as a function of the soil type used in experiment I. Mossoró-RN, 2010

SOIL	PERIOD	ACCUMULATED VOLUME (L $COLUMN^{-1}$)
RQ	06 - 20/08	3,47
RQ	21/08 - 09/09	4,36
PVAe	06 - 20/08	3,30
PVAe	21/08 - 09/09	4,42
CXve	06 - 20/08	4,63
CXve	21/08 - 09/09	5,51
RY	06 - 20/08	3,61
RY	21/08 - 09/09	4,70
VX	06 - 20/08	2,79
VX	21/08 - 09/09	3,16

Table 7. Average values of accumulated irrigation volume for fortnightly periods depending on the crop and soil used in experiment II. Mossoró-RN, 2010

CULTURE	SOIL	PERIOD	ACCUMULATED VOLUME (L $COLUMN^{-1}$)
	PVAe	24/10-07/11	3,30
	CXve	24/10-07/11	2,90
Melao	PVAe	08-23/11	4,54
	CXve	08-23/11	5,29
	PVAe	24-27/11	4,91
	CXve	24-27/11	5,50
	PVAe	24/10-07/11	3,30
	CXve	24/10-07/11	2,90
	PVAe	08-23/11	5,35
Cowpeas	CXve	08-23/11	5,99
	PVAe	24/11-08/12	7,25
	CXve	24/11-08/12	7,86
	PVAe	24/10-07/11	3,30
	CXve	24/10-07/11	2,90
	PVAe	08-23/11	5,15
Corn	CXve	08-23/11	5,44
	PVAe	24-27/11	7,10
	CXve	24-27/11	7,54
	PVAe	24/10-07/11	3,30
	Cxve	24/10-07/11	2,90
	PVAe	08-23/11	5,07
Sorghum	CXve	08-23/11	5,22
	PVAe	24-27/11	6,98
	CXve	24-27/11	7,36
	PVAe	24/10-07/11	7,02
	Cxve	24/10-07/11	3,91
	PVAe	08-23/11	11,9
Pepper	CXve	08-23/11	8,0
	PVAe	24/11-08/12	19,3
	CXve	24/11-08/12	14,9
	PVAe	09-14/12	23,5
	Cxve	09-14/12	18,3

4.5 FEATURES EVALUATED

4.5.1 Soil salinity

During the flowering period of each crop, soil samples were taken to determine soil salinity. In each plot, soil was collected at two depths: 0-20 and 20-40 cm. At each depth, soil samples were taken from three points 10 cm equidistant from the neck of the plant, so that a single sample was taken from the left side, another from the right side and the last from the opposite side of the lateral line at an angle of 90° to the neck of the plant, i.e. in the shape of triangles, forming a composite sample of each soil layer.

The composite samples were air-dried, crumbled and sieved through a 0.002 m mesh, where the electrical conductivity $CE1_{:2.5}$ was determined in the supernatant from the mixture of one part soil to two and a half parts distilled water, according to the methodology proposed by EMBRAPA (1997). The supernatant was obtained by weighing 30 g samples of soil and placing them in a 125 mL flask, then adding 75 mL of distilled water and shaking at 350 rpm for 10 minutes. After stirring, the samples were left to stand for one hour before the EC was measured.

To determine the electrical conductivity of the saturation extract (ECes), a standard measure of soil salinity, calibration curves were made using part of the samples, measuring $EC1_{:2.5}$ and ECes in them. The saturation strata were obtained using saturation paste, carried out according to the methodology described by Richards (1954), using 300g of soil. In the extract, EC was determined using a conductivimeter, Ca^{2}_{+} and Mg^{2}_{+} by titration with EDTA, and Na^{+} using a flame photometer.

Linear regression equations were developed from the $CE12_{.5}$ and CEes data to estimate the CEes data from the CE1:2.5 data, by depth (0-20 and 20-40 cm) and in the soil profile (0-40 cm).

The variables used to assess soil salinity were pH and electrical conductivity of the saturation extract (ECes).

4.5.2 Vegetative parameters

4.5.2.1 Aerial Part Dry Mass

The dry mass of the aerial part of the plants was obtained from the sum of the dry mass of the leaves and stems, which were dried in a forced-circulation oven at a temperature of 65 to 70° C until constant weight. After drying, the material was weighed on a precision scale (0.01g).

4.5.2.2 Growth

Growth was assessed through non-destructive analysis, carried out weekly from the 12th day after sowing (DAS) to 33 (DAS) and destructive analysis (at the end of the experiment) in Experiment I and in Experiment II it was carried out weekly from the 12th DAS to 33DAS for the melon crop, and for the other crops it extended to 40DAS.

Leaf area was obtained from the length and diameter of all the leaves on each plant and estimated using equations proposed for each crop (Table 6). In the last collection, leaf area was estimated using the disc method, which consists of removing leaf discs of known area from four positions on the leaf limb of a set of leaves, distributed symmetrically, avoiding sampling the central vein, according to studies by Huerta (1962) and Gomide *et al.* (1977). The same procedure was used for drying the discs and the leaves from which the leaf discs were removed (forced circulation oven at 65-70° C until constant weight), the mass of which resulted from adding the leaves to the leaf discs, obtaining the total dry matter of the leaves. The area of the disk was correlated with the dry mass of the leaves.

The equations used to estimate the leaf area of the collections are described in Table 8.

Table 8. Equations used to estimate the leaf area of melons, beans, maize, sorghum and peppers. Mossoró, 2010

Culture	Equation	Recommendation
Melao	$\Sigma\ (0.826\ L^{(1)}\text{-}^{89})$	Nascimento *et al.*, 2002
Beans	$\Sigma\ (0.9915 \times (L \times W)^{(0.9134)})$	Lima *et al.*, 2008

Corn	Σ ((L x W) x 0.75)	Montgomery, 1911
Sorghum	Σ ((L x W) x 0.75)	Montgomery, 1911
Pepper	Σ ((L x W) x 0.60)	Trivelli *et al.*, 1997

4.5.3 Gas exchange analysis

Stomatal conductance was measured in Experiment I at 33 days after sowing, using a model AP-4 dynamic diffusion meter (Delta-T devices). Daily stomatal conductance curves were drawn up by reading the stomatal conductance of a well-lit leaf. Together with the first reading, the leaves were marked and subsequent readings were taken on the same leaves and in the same initial sequence. A fully expanded leaf with no symptoms of damage was assessed. The sum of all the readings taken during the period resulted in the accumulated conductance. The readings were taken between 06:00 and 16:00 on a day with good sunshine.

4.5.4 Mineral element content in leaves

The levels of mineral elements in the dry mass of the leaves from Experiment II were determined according to the methodology of Embrapa (2009), with N obtained by digesting 0.2 g of the sample in 3 ml of H_2SO_4; 1 ml of $H_2O_{(2)}$and 1 g of digester mixture in the ratio 10:1, where 10 is $CuSO_4$ and 1 is K_2SO_4, and the other nutrients were determined using nitric digestion, where 0.5 g of the sample was digested in 5 ml of HNO_3, where P was determined by colorimetry, Na and K by flame photometry, Ca, Mg, Cu, Zn, Fe and Mn by atomic absorption spectrophotometry.

4.5.5 Statistical analysis of data

The data obtained was subjected to analysis of variance and the F test using software SAEG 9.0 (RIBEIRO JUNIOR, 2001). As the salinity data are quantitative variables, regression analysis was carried out, where the coefficients of the models were tested based on the mean square of the residual from the analysis of variance, according to Ribeiro Junior (2001). The soils were compared using the Tukey test at 5% probability and regression analysis.

5. RESULTS AND DISCUSSION

CHARACTERISTICS EVALUATED IN EXPERIMENT I

5.1 SOIL SALINITY EVALUATION

5.1.1 Ratio between the electrical conductivity of the saturation extract (ECes) and the electrical conductivity of the soil-water suspension 1:2.5 ($EC_{1:2.5}$)

Soil salinity is assessed by the electrical conductivity of the saturation extract. However, due to the slowness of the laboratory method, an alternative is to estimate from the diluted extract (DANTAS *et al*, 2005; RHOADES and MIYAMOTO, 1990), but this requires calibration between the methods before it can be used.

Figures 3, 4, 5, 6 and 7 show the relationship between $EC1_{:2.5}$ and ECes at the depths of 0-20 cm, 20-40 cm and in the profile (0-40 cm) of the soils studied. A linear relationship can be observed with a high degree of coefficient of determination in both layers and in all the soils studied. In the Quartizarenic Neosol (RQ), the concentration factor ranged from 7.38 to 8.03, with an average concentration value in the profile of 7.74 (Figure 9). For the Eutrophic Yellow Red Argisol (PVAe), the concentration factor varied from 7.66 to 9.58 and the average value in the profile was 8.87 (Figure 10). The variation found for the Cambissolo Hàplico (CXve) was from 5.49 to 6.18 and the average value in the profile was 5.93 (Figure 11). In the Fluvial Neosol (RY), the concentration factor varied from 4.49 to 5.10 and the average value in the profile was 4.87 (Figure 12). In the

In the Vertissolo (VX), the concentration factor ranged from 4.09 to 3.51 and the profile average was 3.71 (Figure 7). There was the least variation in the concentration factor in VX (0.58) and the greatest variation in PVAe (1.92). However, in all the soils studied, the ECes can be estimated using the equations obtained from the average values in the profile, with good approximation of the data. The results obtained in this work confirm the data obtained by Dantas *et al*. (2005), who found that ECes can be estimated from $EC_{I;2}$ and observed that in both methodologies the joint analysis of the two depths can be used to make this estimate.

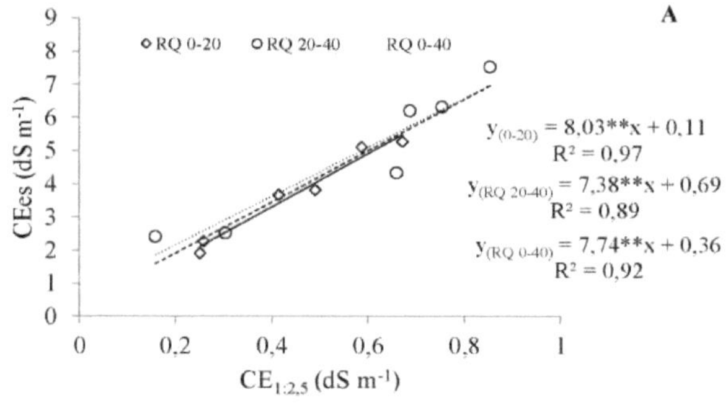

Figura 9. Relationship between $EC_{1:2.5}$ and ECes at the depths of 0-20 cm, 20-40 cm and in the profile (0-40 cm) in Quartizarenic Neosol (RQ) cultivated with melon until flowering. Mossoró, 2010

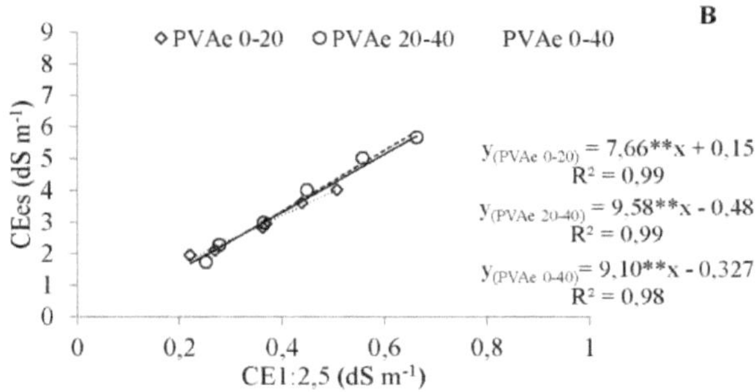

Figura 10. Relationship between EC12.5 and ECes at the depths of 0-20 cm, 20-40 cm and in the profile (0-40 cm) in Eutrophic Yellow Red Argisol (PVAe) cultivated with melons until flowering. Mossoró, 2010

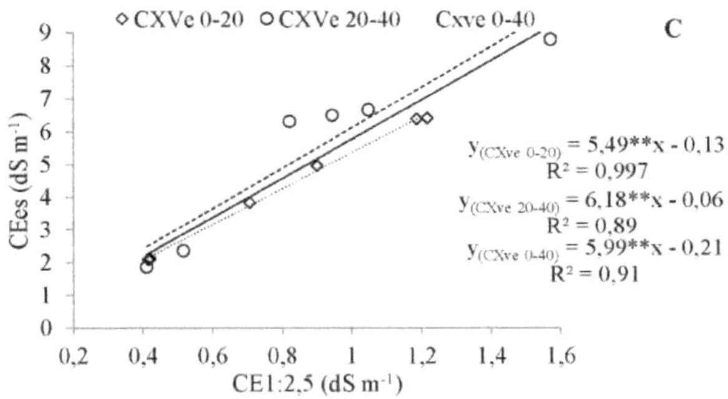

Figura 11. Relationship between EC12.5 and ECes at depths of 0-20 cm, 20-40 cm and in the profile (0-40

cm) at Cambissolo Hàplico (CXVe) cultivated with melons until flowering. Mossoró, 2010

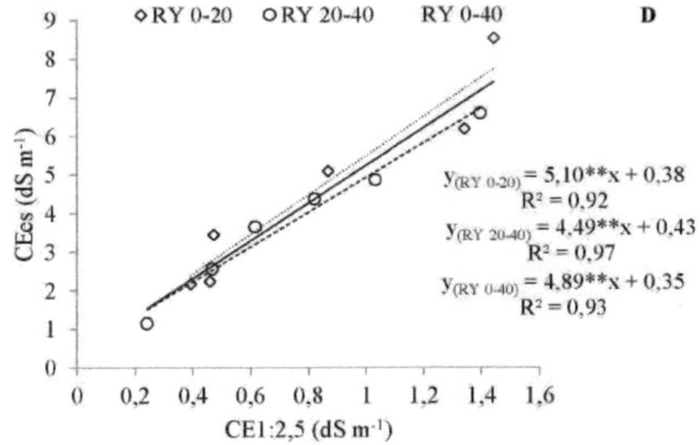

Figura 12. Relationship between $EC_{1:2.5}$ and ECes at the depths of 0-20 cm, 20-40 cm and in the profile (0-40 cm) in Fluvial Neosol (RY) cultivated with melon until the flowering period. Mossoró, 2010

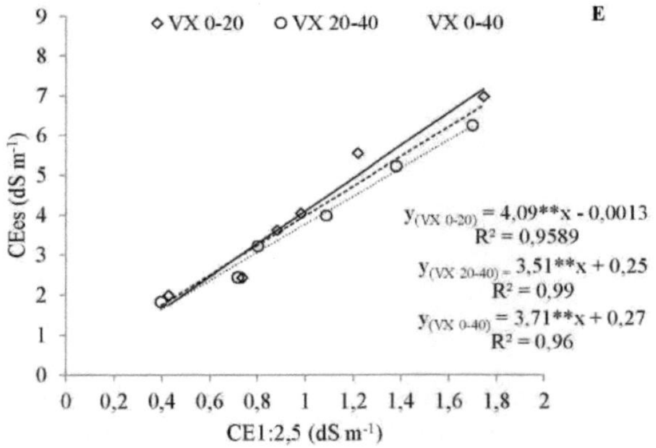

Figura 13. Relationship between $EC_{12.5}$ and ECes at depths of 0-20 cm, 20-40 cm and in the profile (0-40 cm) in Vertissolo (VX) cultivated with melon trees until flowering. Mossoró, 2010

Figure 14 shows the EC values in the root profile as a function of the water table for the soils studied. In all the soils, there was an increase in EC as a result of the increase in water table. However, this behavior differed between the soils, with a greater tendency for salinity to increase in CXVe, RY and VX, due to the higher content of the clay fraction, which limits the leaching of salts as a result of the leaching rate applied. In NQ and NY, salts are leached into the soil profile because the texture is sandier - 930 and 935 g kg^{-1} of sand, respectively. The EC is directly proportional to the ECa, a behavior that confirms the theory of Ayers and Westcot (1999) and Medeiros *et al.* (1998).

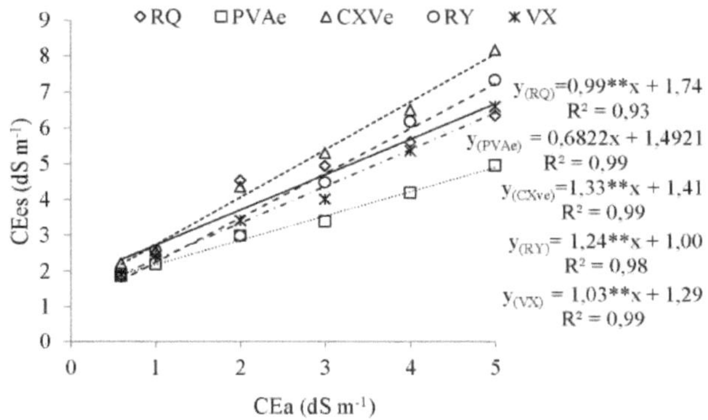

Figura 14. Relationship between ECes in the soil profile (0-40 cm depth) estimated by CE1:5 as a function of ECa in soils of the Mossoró-Assu agropolis cultivated with melon trees until the flowering period. Mossoró, 2010

For all the soils studied, ECes was higher than ECa for the water with the lowest salinity level A1 (0.59 dS m^{-1}), and in A2 (1.0 dS m^{-1}) for RQ, PVAe, RY and VX. In CXVe, ECes increased considerably from ECa (1.0 dS m^{-1}), which suggests greater soil salinization problems due to the higher content of the clay fraction. RY was the soil in which the highest EC values were observed at the highest salinity level. This soil has a higher silt content (360 g $kg^{-1)}$ than the other soils studied, which implied a lower leaching fraction. The expected leaching rate was 10%, but the low hydraulic conductivity of RY and VX meant that this rate could not be fully applied.

5.1.2 Soil salinity and pH profiles

Figure 15 shows that salinity in the soil profile showed little variation along the profile for all the soils studied, which can be explained by the high leaching fraction applied, as explained by Ayers and Westcot (1999).

In soils with a sandier texture, such as RQ and PVAe, there was a tendency for salts to accumulate more at a depth of 40 cm. This behavior can be explained by the higher content of the sand fraction in these types of soils, causing salts to be leached into the profile, This behavior is different from that observed for CXve, RY and VX, which showed a greater accumulation of salts on the surface when compared to the above-mentioned soils, remaining constant throughout the profile, which can be explained by the reduced drainage capacity of these types of soils (Figure 15).

The pH values were not significantly influenced by the salinity of the irrigation water, remaining constant and with close values between the soils studied.

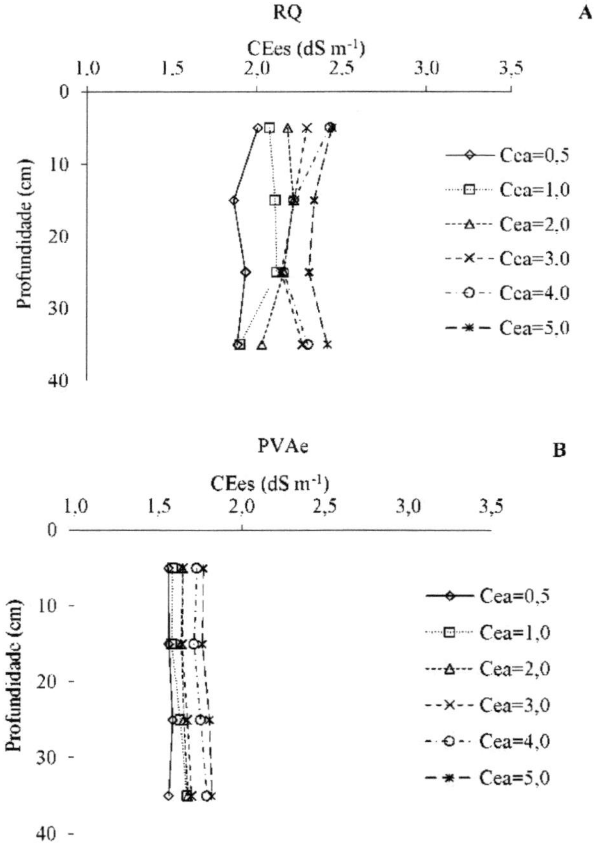

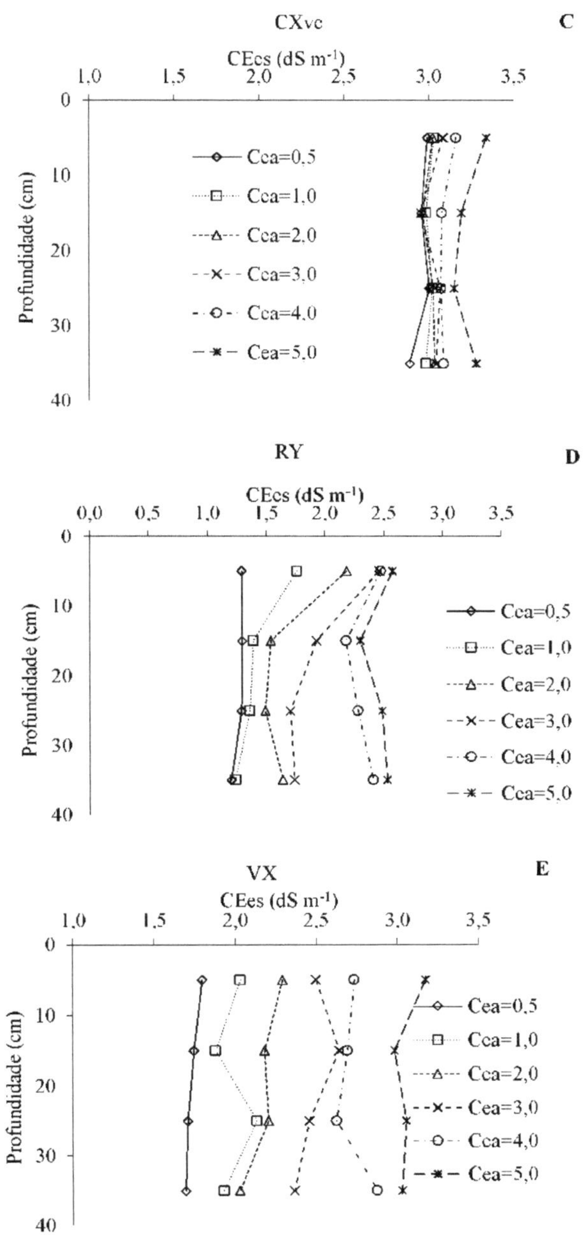

Figura 15. ECE profiles estimated from soil CEi:2.5 as a function of irrigation water salinity in RQ (A), PVAe (B), CXve (C), RY (D) and VX (E) at 36 days of melon cultivation. Mossoró, 2010

It was observed that the salinity values in all the soils studied grew linearly as a result of the increase in salinity of the irrigation water, at 15 and 36 days of cultivation (Figure 16). In RQ and PVAe, due to their sandier constitution, which is more efficient at leaching, the concentration of salts was lower, which can be seen from the angular and linear coefficients of the regression equations (Figure 16A). On the other hand, in CXve, RY and VX this increase was more evident, especially at 36 DAS. This can be explained by the higher clay content in this type of soil, which decreases leaching efficiency, promoting a greater accumulation of salts in the soil (Figure 16B). In both soils, salinity levels could be estimated at 15 and 36 DAS by the equations with an accuracy of over 84%. The values obtained in this study for CXve, RY and VX agree with Santana *et al.* (2006), who obtained higher EC values in the clay-textured soil, with 2.26, 4.86, 6.89 and 10.25 dS m^{-1} for water of 0.1, 2.0, 5.0 and 8.0 dS m^{-1}, respectively. This result was also observed by Souza (1995), who studied the evolution of ECes and found values proportional to the treatments of excess salt in the irrigation water, in a soil cultivated with bean plants.

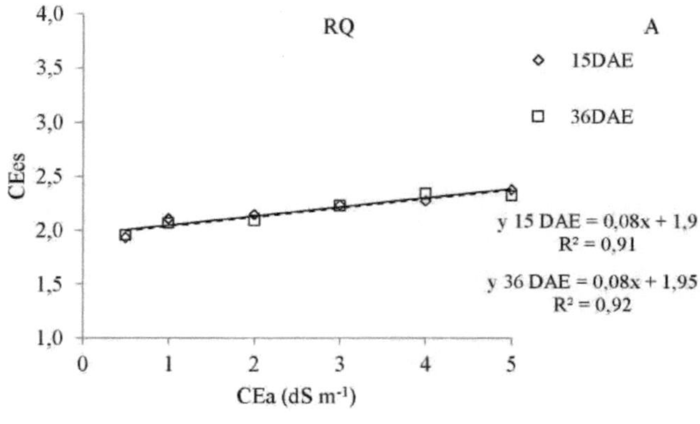

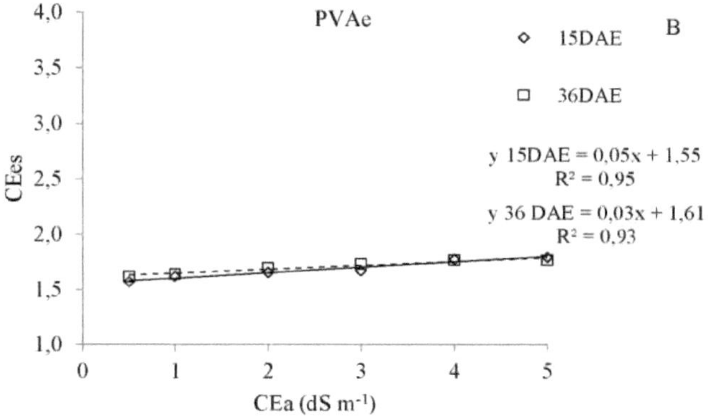
PVAe
B
15DAE
36DAE
y 15DAE = 0,05x + 1,55
R² = 0,95
y 36 DAE = 0,03x + 1,61
R² = 0,93
CEes
CEa (dS m-1)

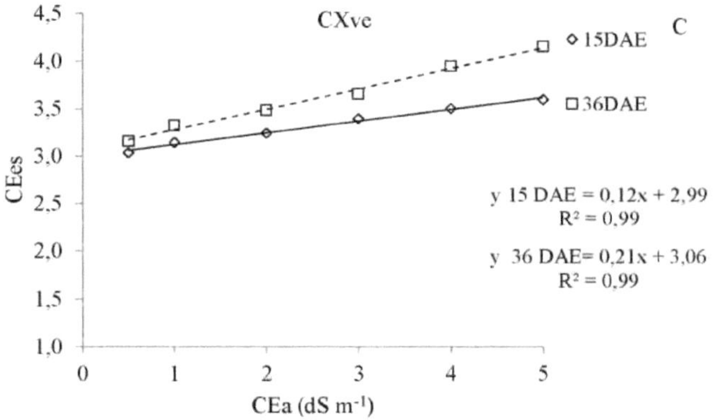
CXve
C
15DAE
36DAE
y 15 DAE = 0,12x + 2,99
R² = 0,99
y 36 DAE= 0,21x + 3,06
R² = 0,99
CEes
CEa (dS m-1)

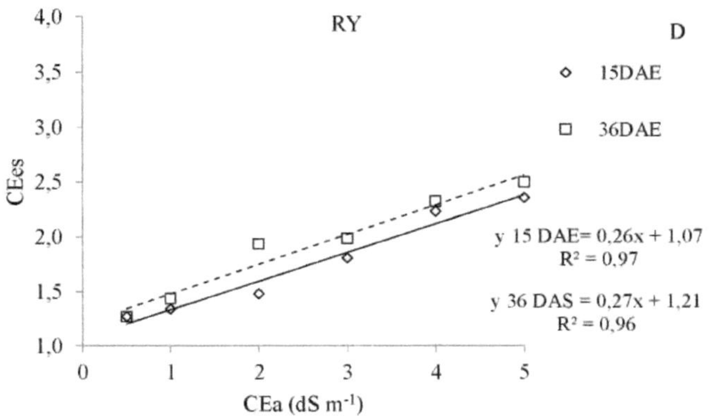
RY
D
15DAE
36DAE
y 15 DAE= 0,26x + 1,07
R² = 0,97
y 36 DAS = 0,27x + 1,21
R² = 0,96
CEes
CEa (dS m-1)

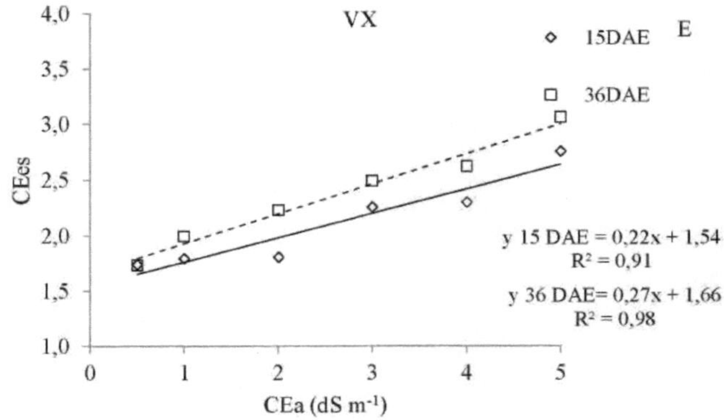

Figura 16. Evolution of ECes in the soil profile as a function of irrigation water salinity at 15 and 36 DAE in RQ (A), PVAe (B), CXve (C), RY (D) and VX (E) grown with melon until flowering. Mossoró, 2010

5.2 EVALUATION OF VEGETATIVE PARAMETERS

5.2.1 Aerial Part Dry Matter

For the total dry matter of the melon tree, there was no significant interaction between the soils and the salinity of the irrigation water, with only isolated effects for the soil and water factors (Table 9).

Table 9. F-test values and statistical significance for the dry matter content of the melon tree under salt stress in soils of the Mossoró-Assu agropolis. Mossoró, 2010

FV	GL	MSA	MSR
Soil	4	50,845**	9,957**
Water	5	10,687**	10,577**
S x A	20	1.163^{ns}	1.062^{ns}
Waste	60		
CV (%)		15,01	15,76

**significant at 1% probability,*significant at 5% probability and ^{ns} not significant

The highest total dry matter values were observed in VX, followed by CXve and RY, which did not differ statistically, and the lowest values in RQ (Table

10). These results suggest that even under saline stress, soils with a loamier texture make more nutrients available to the plant, due to their greater water and nutrient retention capacity.

Table 10. Average dry matter values of the aerial part of the melon tree under saline stress in representative soils of the Mossoró-Assu agropolis. Mossoró, 2010

SOILS	EC IRRIGATION WATER (dS m^{-1})						Medium
	0,59	1,0	2,0	3,0	4,0	5,0	
RQ	7,12 C	7,59 B	6,66 B	6,20 B	4,97 B	4,62 B	6,19 C
PVAe	8.41 BC	8,15 B	6,60 B	7,69 B	5,69 B	6,57 B	7.45 BC
CXve	11,17 A	9.12 AB	8,68 B	7,01 B	6,46 B	6,36 B	8,13 B
RY	9.81 AB	8,46 B	8,35 B	7,89 B	6,98 B	6,41 B	7,98 B
VX	11,75 A	11,58 A	12,28 A	11,93 A	11,90 A	10,67 A	11,69 A

Values followed by the same letter in the column do not differ by Tukey's test at 5% probability.

There were decreasing linear model fits for RQ, PVAe, CXve and RY, while there was no model fit for VX. From the lowest level (0.59dS m^{-1}), there were already decreases in the total dry matter of the plants, showing the effect of salinity, with reductions of 0.66; 0.50; 1.01; 0.66 g $plant^{-1}$ per unit increase in salinity, resulting in estimated losses of 43.95; 38.80; 31.44; 26.72% for CXve, RQ, RY and PVAe, respectively (Figure 17).

Reduction in dry matter has been a classic behavior verified by several authors when plants are subjected to saline stress, including Meloni *et al*, (2001); Botia *et al*, (2005). This behavior has been attributed to the reduction in the water potential of the soil solution generated by the osmotic effect of the added Na^+ and Cl^- salts, making it difficult for the plant roots to absorb water (MARSCHNER, 1995; BRAY *et al*, 2000).

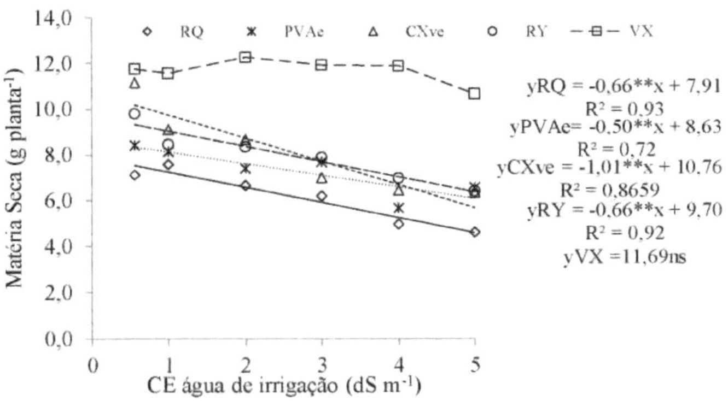

Figure 17. Dry matter of the aerial part of melon trees grown until flowering as a function of ECa in soils of the Mossoró-Assu agropolis. Mossoró, 2010.

**significant at 1% probability,*significant at 5% probability and [ns] not significant

5.2.2 Leaf Area

There was a significant effect on the leaf area of the melon tree for the interaction between soil types and levels of salinity in the irrigation water (Table 11). With the exception of RQ and PVAe, there were decreases in leaf area values. The non-effect of RQ and PVAe was probably due to the lower accumulation of salts, a consequence of the greater leaching capacity of these types of soils, thus reducing the effect of salinity on this characteristic (Figure 12).

Table 11. F-test values and statistical significance for the leaf area of the melon tree under salt stress in soils of the Mossoró-Assu agropolis at four collection times. Mossoró, 2010

FV	GL	AF
Soil	4	95,217**
Water	5	6,72**
Soil x Water	20	1,85*
Error (A)	58	
Season	3	4667,5**
Soil x Season	12	5,65**
Water x Season	15	1.32^{ns}
Soil x Water x Season	60	0.98^{ns}
Error (B)	182	
CV (%)		4,89

**significant at 1% probability,*significant at 5% probability and ^{ns} not significant

Considering the soils where there were significant effects of irrigation water salinity, linear decreases were observed in CXve and VX as a function of increasing irrigation water salinity, while in RY a 3rd degree polynomial model was fitted (Figure 18).

According to Terry and Waldron (1984) one of the first responses of plants subjected to saline stress is a reduction in the rate of leaf growth, reducing the leaf area available for photosynthesis. In general, it can be seen that the higher the level of saline water, the smaller the leaf area. This behavior can be attributed to the increase in soil salinity with growing time, thereby reducing the osmotic potential of the soil solution, decreasing the availability of water to plants and causing crops to suffer a progressive reduction in growth and production whenever the saline concentration increases.

Franco *et al.* (1997) found that leaf area decreased with salinity for different melon cultivars, although the effects differed between cultivars. Folegatti and Blanco (2000) found similar results in cucumber, with a reduction in leaf area as the salinity of the irrigation water increased.

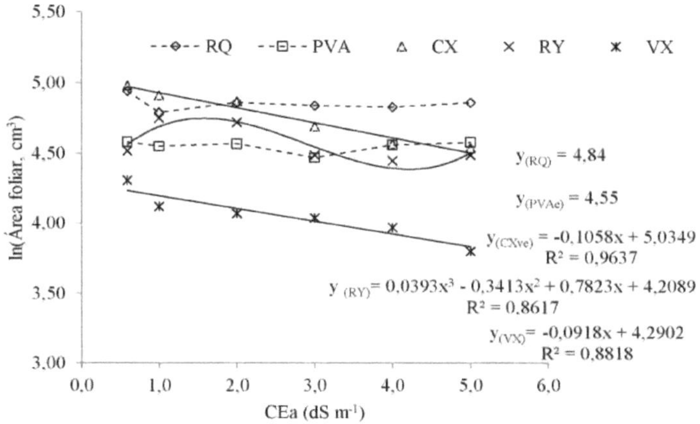

Figure 18. Leaf area of the melon tree as a function of the salinity of the irrigation water in soils of the Mossoró-Assu agropolis (Average of the seasons). Mossoró-RN, 2010

The average leaf area values at the lowest salinity level (0.59 dS m^{-1}) were statistically similar for RQ and CXve, which were superior to PVAe, RY and VX, which did not differ from each other. For the 1.0 dS m^{-1} level, the highest leaf area values were obtained in RQ, CXve and RY, which did not differ statistically, followed by PVAe and VX, the latter being the soil with the lowest value for this variable. At the 2.0 dS m^{-1} level, similar behavior was observed for all the soils studied, except for VX, where the lowest leaf area value was observed. At the fourth salinity level, similar behavior was observed for RQ and CXve, which did not differ statistically from PVAe and RY, and lower values were observed in VX. For salinity of 5.0 dS m^{-1}, RQ, PVAe and CXve showed higher values and were similar to each other, followed by RY and VX, the latter showing the lowest average value (Table 12).

It was observed that as the salinity of the irrigation water increased, the soils most affected by salinity were the clayey ones, in which the lowest leaf area values were observed. This may have been due to the accumulation of salts on the surface, since these soils have restricted drainage, thus interfering with water absorption by the plant as a result of the decrease in the osmotic potential of the water in the soil caused by the excess salts.

Table 12. Leaf area of melon as a function of soil type for each level of irrigation water salinity. Mossoró, 2012

SOILS	EC IRRIGATION WATER (dS m^{-1})					
	0,59	1,0	2,0	3,0	4,0	5,0
RQ	4,94 A	4.79 AB	4,86 A	4.84 AB	4.83 AB	4.86 AB
PVAe	4,58 B	4,55 B	4,57 A	4,47 B	4.56 AB	4.58 AB
CXve	4,98 A	4.91 AB	4,87 A	4.69 AB	4.57 AB	4,54 B
RY	4,52 B	4.75 AB	4,72 A	4,49 B	4,45 B	4,49 B
VX	4,31 B	4,12 C	4,07 B	4,04 C	3,97 C	3,80 C

Values followed by the same letter in the columns do not differ by the Tukey test at 5% probability.

*Values expressed as ln (leaf area, in cm^2)

The leaf area grew according to a 2nd degree polynomial in all the soils studied, with the growth rate increasing over time. This is common for phases I and II of crops. The highest values of this variable were observed at the end of the evaluation period in RQ and the lowest values in VX (Figure 19).

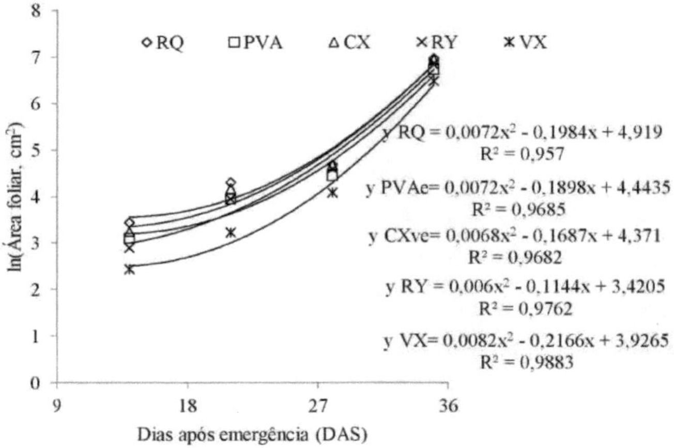

Figura 19. Leaf area of the melon tree as a function of plant age in soils of the Mossoró-Assu agropolis. Mossoró-RN, 2010

In the first growing season (9 DAS), the highest average leaf area values were observed in RQ, which was similar to PVAe and Cxve, which did not differ from RY and the lowest value for VX. At 18 DAS, the highest leaf area averages were obtained in RQ and CXve, followed by PVAe and RY, with the lowest value observed in VX. In the third and fourth evaluation periods (27 and 36 DAS), RQ, CXve and RY were similar and did not differ from PVAe, with VX being the soil in which the lowest leaf area values were observed (Table 13).

In all the evaluation periods, there was a tendency for the clayier soils VX, RY and CXve to show lower leaf area values, emphasizing the effect of salinity in soils with a clayier texture, which can be explained by the high ECes and restricted drainage, resulting in lower leaching efficiency.

Table 13. Leaf area of the melon tree as a function of soil type at each growth analysis time. Mossoró-RN, 2010

	ANALYSIS PERIODS			
SOILS	9 DAE	18 DAE	27 DAE	36 DAE
RQ	3.44 AB	4,31 A	4.68 AB	6.97 AB
PVAe	3.09 BC	3,95 B	4,45 B	6,72 B
CXve	3.24 ABC	4,17 A	4.68 AB	6.95 AB
RY	2,89 C	3,95 B	4.60 AB	6.83 AB
VX	2,43 D	3,22 C	4,08 C	6,48 C

Values followed by the same letter in the columns do not differ by the Tukey test at 5% probability.

5.3 Stomatal Conductance

Figure 14 shows the variation in stomatal conductance (SG) during the initial growth of the melon tree for the different types of soil subjected to different levels of salinity. In general, it can be seen that the maximum conductance peaks for the different soil types occur mostly at 8:00 am, due to the low intensity of radiation at this time and daytime irrigation. These values then drop sharply at 12 noon for the RQ, CXve and RY soils due to the lower soil humidity at this time.

It can be seen that there was a reduction in the opening of stomata in the afternoon (16:00) in all the soils evaluated, even with high light intensity, especially in plants irrigated with higher ECa levels. This may be due to the melon tree's water regulation mechanism when subjected to low soil water potentials and/or high evapotranspiration demand.

Conductance was reduced in the plants in the RQ, PVAe, RY and VX soils by the increase in salinity of the irrigation water (Figure 20A), when compared to the lower salinity water. This suggests that there was a reduction in the flow of water from the plant into the system, probably due to the concentration of salts in the root zone, reducing plant metabolism. Similar results of a decrease in transpiration rate due to the addition of NaCl were also found in *Cucumis melo* (SOARES *et al.*, 2002) and *Psidium guayava* (TAVORA *et al.*, 2001), corroborating the results of this study. It can also be seen that the plants subjected to a conductivity of 0.59 dS m^{-1} showed stomatal conductance values that were mostly higher than the treatments subjected to salt stress. Working with Conilon coffee rootstocks, Silva *et al.* (2010) found a significant reduction in stomatal conductance in response to water deficit.

Evaluating stomatal conductance in each type of soil, it can be seen that there was a greater effect of salinity in the RQ, CXve and VX soils, with this effect being more evident in the readings taken in the morning (FIGURES 20A, 20C and 20D). In VX, the plants irrigated with the lowest salinity water showed the highest values throughout the day (FIGURE 20E). This behavior can be attributed to the physico-chemical characteristics of VX, since it has a high clay content and a low leaching rate. Therefore, plants grown in this type of soil were more sensitive to salt stress.

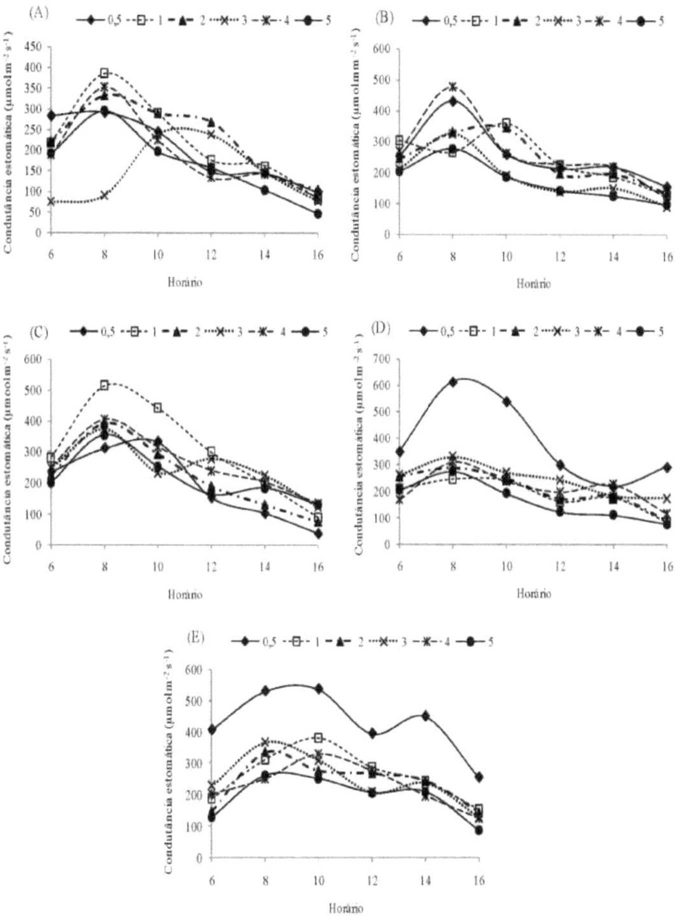

Figura 20. Stomatal conductance in the initial phase of the melon tree grown in different types of soil (A - RQ), (B - PVAe), (C - CXve), (D - RY) and (E - VX) and irrigated with water of different salinity levels at 35 days after emergence.

CHARACTERISTICS EVALUATED IN EXPERIMENT II

5.4 Aerial Dry Matter

There was a significant effect of the interaction between the soil and water factors for the dry matter of the aerial parts of the bean and sorghum crops, an isolated effect of the soil for the melon and maize crops, and of the salinity levels of the irrigation water for the maize crop and for the pepper, which was not significant (Table 14). 1

Table 14. F-test values and statistical significance for the dry matter content of melon, bean, maize, sorghum and pepper crops under saline stress in PVAe and CXve. Mossoró, 2010

FV	Melao	Beans	Corn	Sorghum	Pepper
Soil	22,354**	1.407^{ns}	4,763*	29,207**	3.773^{ns}
Water	2.824^{ns}	133,476**	22,663**	84,125**	1.819^{ns}
S x A	1.3685^{ns}	12,74**	3.262^{ns}	18,35**	0.196^{ns}
CV (%)	14,278	6,418	9,790	5,789	18.599^{ns}

**significant at 1% probability,*significant at 5% probability and ^{ns} not significant

The melon tree showed the highest average dry matter value in CXve (Figure 15). For the bean crop in the PVAe, there was a reduction in dry matter content only for the highest salinity level, but in the CXve there was a reduction at all salinity levels (Table 15). Comparing the soils within each salinity level, it can be seen that for the ECa of 0.59 dSm^{-1}, the highest total dry matter values were obtained in CXve, and for the other salinity levels the opposite situation occurred, i.e. the highest averages were obtained in PVAe (Table 15).

In the sorghum crop, there was a reduction in dry matter in the PVAe only for the highest salinity level (5.0 dS m^{-1}), while in the CXve there was a reduction from the second salinity level (2.75dS m$^{-1)}$, which was statistically similar to the highest salinity level. Analyzing the soils within each salinity level, there was a statistical similarity for the lowest salinity level and in the other levels the highest averages were obtained in the PVAe (Table 15).

For the melon and corn crops, the highest average values were obtained in CXve and PVAe, respectively. For the salinity levels, there were decreases in dry matter in the maize crop from the second level (2.75dS m^{-1}), which was statistically similar to the most saline level, with a total reduction of 30.54% between the highest and lowest salinity levels (Table 16).

For the pepper crop, there was no significant influence of irrigation water salinity on dry matter content in the two soils studied.

Several authors have also found a reduction in the dry matter of different crops as a result of increased salinity in the medium, including Lima *et al*. (2007) and Oliveira *et al.* (2009).

According to Rhoades *et al.* (1992), excess salts reduce plant development due to the increase in energy that needs to be released to absorb water from the soil and the biochemical adjustment required to survive under stress.

Table 15. Average dry matter values of the aerial part of different crops under saline stress in PVAe and CXve. Mossoró, 2010

Cultures	Soil	EC IRRIGATION WATER (dS m 1)		
		0,59	2,75	5,00
Beans	PVAe	20.38 b A	19.37 a A	13.08 a B
	CXve	23.46 a A	17.36 b B	10.15 b C
Sorghum	PVAe	17.23 a A	16.08 a A	13.12 a B
	CXve	18.35 a A	11.55 b B	10.15 b B

Values followed by the same lower-case letter in the column and upper-case letter in the row do not differ by the Tukey test at 5% probability.

Table 16. Average dry matter values of the aerial parts of melon, corn and pepper as a function of soil type and ECa. Mossoró, 2010

Soils	Cultures		
	Melao	Corn	Pepper
PVAe	9,47 B	16,77 A	2,10 A
CXve	13,05 A	15,16 B	1,77 A
ECa			
0,59	12,32 A	19,32 A	2,07 A
2,75	11,33 A	15,15 B	2,03 A
5,00	10,12 A	13,42 B	1,71 A

Values followed by the same letter do not differ by Tukey's test at 5% probability.

5.5 Leaf Area

For the leaf area of the melon tree, there was a significant interaction between the different salinity levels and the evaluation times (Table 17).

Splitting the salinity levels within each evaluation time, there was a statistical difference only at 33 DAE, with decreases in AF as salinity increased, ranging from 437.32 to 648.41 cm^2 for the ECa of 5.00 and 0.59 dS $^{m(-1)}$, respectively (Table 18).

The development of the melon tree was characterized by slow initial growth until 26 DAE at the different salinity levels (Figure 21). In the melon crop, the high concentration of salts in the soil and water cause morphological, structural and metabolic changes and inhibit its growth and development, reducing the percentage of dry mass of the plants (SIVRITEPE *et al., 2003*), 2003), showing that the higher the salinity of the irrigation water, the smaller the leaf area, making this decrease evident over time due to the accumulation of salts in the soil, making the washing out of salts in the soil an efficient management practice.

Table 17. F-test values and statistical significance for the initial growth of the melon tree under saline stress in soils from the Mossoró-Assu agropolis at four collection times. Mossoró, 2010

FV	GL	AF
Soil	1	3.94^{ns}
Water	2	4,54*
Soil x Water	2	1.27^{ns}
Error (A)	12	
Season	3	275,08**
Soil x Season	6	0.47^{ns}
Water x Season	3	2,89*
Soil x Water x Season	6	1.15^{ns}
Error (B)	36	
CV (%)		33,9

**significant at 1% probability,*significant at 5% probability and ^{ns} not significant

Table 18. Leaf area of the melon tree as a function of irrigation water salinity at each growth analysis time. Mossoró-RN, 2010

Levels of	ANALYSIS PERIODS			
Salinity	12 DAE	19 DAE	26 DAE	33 DAE
0,59	22,28 a	62,3 a	140,81 a	648,41 a
2,75	17,18 a	65,54a	112,91a	536 b
5,00	18,88a	39,56 a	97,65 a	457,32 c

Values followed by the same letter in the columns do not differ by the Tukey test at 5% probability.

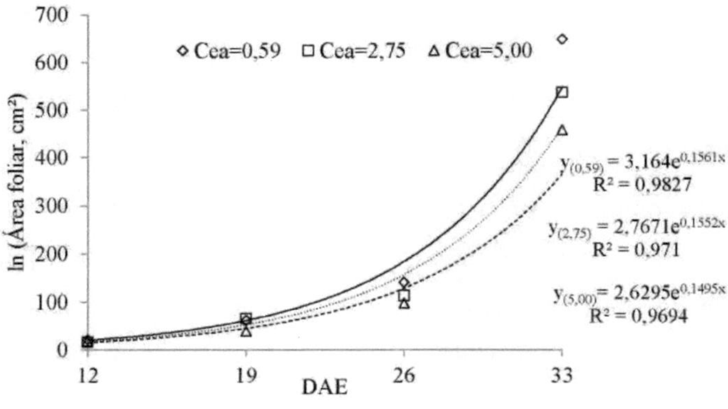

Figure 21. Leaf area of the melon tree under saline stress at different evaluation times. Mossoró-RN, 2010

In the bean crop, there was a significant effect of the soil-water-season interaction on PA (Table 15). When the water-soil interaction was broken down, there were decreases in PA in the PVAe as a function of the increase in irrigation water salinity, while in the CXve the highest values were observed at the lowest salinity level and the others were similar. At 0.59 dS m^{-1} the PA was similar in both soils, at 2.75 dS m^{-1} it was higher in PVAe and at 5.00dS m^{-1} the highest values were observed in CXve (Table 20). Splitting the salinity levels within each evaluation time, there was a significant difference between the ECa from 33 DAE onwards, when there was a reduction in PA as the salinity of the irrigation water increased (Table 21). There was no mathematical model to explain the biological behavior of the leaf area of the bean as a function of salt stress.

Leaf area values in the bean plant increased until 33 DAE at all salinity levels, with reductions at 40 DAE. This behavior can be explained by possible leaf senescence due to a change in the plant's stage of development. Reductions in bean leaf area under salt stress were also observed by O'Learly (1971),

Sousa (1995) and Silva (1998).

Table 19. F-test values and statistical significance for leaf area of bean, maize, sorghum and pepper under saline stress in soils of the Mossoró-Assu agropolis at five collection times. Mossoró, 2010

FV	GL	Beans	Corn	Sorghum	Pepper
Soil	1	0.1^{ns}	0,02**	24,1**	34.82^{ns}
Water	2	75,31**	45,29**	3,26*	151,77**
S x A	2	8,05**	9,48**	0.48^{ns}	26,31**
Error (A)	12				
Season	4	2848,58**	8508,91**	372,36**	161,96**
S x E	8	1.16^{ns}	82,32**	1.66^{ns}	96,99**
A x E	4	50,71**	51,9**	0.85^{ns}	31,37**
S x A x E	8	5,54**	10,82**	0,32**	25,57**
Error (B)	48				
CV (%)		10,6	4,1	18,4	13,90

**significant at 1% probability,*significant at 5% probability and ^{ns} not significant

Table 20. Average values of leaf area in bean, corn and pepper crops under saline stress in PVAe and CXve. Mossoró, 2010

Cultures	Soil	EC IRRIGATION WATER (dS m^{-1})		
		0,59	2,75	5,00
Beans	PVAe	2075.91 Aa	1823.75 Ba	1344.55 Cb
	CXve	2021.85Aa	1684.95 Bb	1573.95 Ba
Corn	PVAe	877.54 Aa	885.1 Aa	747.86 Bb

	CXve	907.13 Aa	819.9 Bb	788.22 Ba
Pepper	PVAe	317.87 Ab	273.77 Ba	242.89 Ca
	CXve	462.92 Aa	253.22 Ba	178.92 Cb

Values followed by the same capital letter in the row and small letter in the column do not differ by the Tukey test at 5% probability.

Table 21. Bean leaf area as a function of irrigation water salinity at each growth analysis time. Mossoró-RN, 2010

EC	EVALUATION PERIODS				
(dS m-1)	12 DAE	19 DAE	26 DAE	33 DAE	40 DAE
0,59	240,6a	573,7a	1105,4a	7201,9a	1122,8 a
2,75	238,5a	616,1a	1086,1a	5782,6b	1048,4 a
5,00	256,7a	558,8a	913,8 a	4701,8c	847,2 b

Values followed by the same letter in the columns do not differ by the Tukey test at 5% probability.

For the leaf area of the corn crop, there was a significant effect of the soil-water-season interaction (Table 19). In the PVAe, the highest values were observed at 0.59 and 2.75 dS m^{-1} and the lowest values at 5.00 dS m^{-1}. For the CXve, statistically similar values were observed at 2.75 and 5.00 dS m^{-1} and lower than 0.59 dS m^{-1}. Considering the salinity levels, at 0.59 dSm^{-1} the values were statistically similar for the two soils, while at EC 2.75 dS m^{-1}, the highest value was in the PVAe and for 5.00 dS $m^{(-1)}$) in the CXve (Table 20).It was observed that the maize plants showed exponential growth up to 40 DAE regardless of the salinity level (Figure 22).

Totawat and Mehta (1985), working with maize and sorghum genotypes, found that the PA of both species was negatively influenced by salt stress. At 12.0 dS m^{-1} of electrical conductivity, maize and sorghum plants had their PA reduced by 46 and 60 %, respectively.

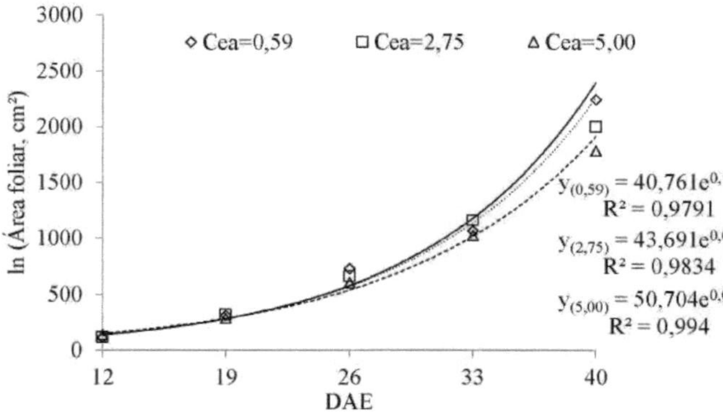

Figura 22. Corn leaf area under saline stress at different evaluation times. Mossoró-RN, 2010

For the sorghum crop, there was an isolated effect between soils, salinity levels and collection times (Table 19). The highest average leaf area values were observed in CXve. In relation to salinity levels, there were decreases in leaf area as salinity increased (Table 22). The sorghum crop showed exponential leaf area growth as a function of the evaluation periods (Figure 23).

Table 22. Average leaf area values of sorghum grown in PVAe and CXve under salt stress. Mossoró, 2010

SOILS		
PVAe	CXve	
883,74 B	1159,42 A	
EC IRRIGATION WATER (dS m ı)		
0,59	2,75	5,00
1108,094 A	1024,156 AB	932,48 B

Values followed by the same letter do not differ by Tukey's test at 5% probability.

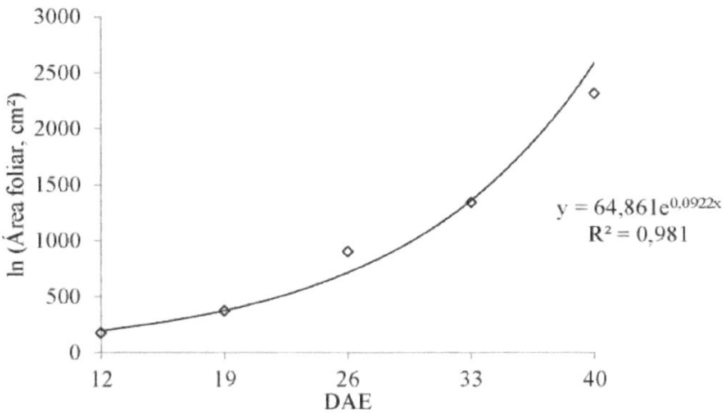

Figure 23. Leaf area of the sorghum crop as a function of each season of growth analysis. Mossoró- RN, 2010

A significant effect was observed for the soil-water interaction for the leaf area of the pepper crop (Table 19). If the soil-water interaction is broken down, it can be seen that for both types of soil there was a reduction in leaf area as salinity increased. By unfolding the soils for each salinity level, it was observed that at 0.59 dS m^{-1} the CXve showed the highest value, at 2.75 dS m^{-1} there was no significant difference, while at 5.00 dS m^{-1} the PVAe showed the highest average value (Table 20).

When the soils were divided into each evaluation period, it can be seen that at 12 DAE there were no differences between the soil types. At 19 and 26 DAE, PVAe had higher average values, while at 33 and 40 DAE the opposite situation occurred, i.e. CXve provided greater increases in PA (Table 23).

By splitting the evaluation periods for each soil type, exponential growth was observed in both soils. Although CXve had the lowest AF up to 33 DAE, this soil showed the fastest growth rate during the evaluation periods, reaching the highest average value at 40 DAE (Figure 24).

Table 23. Pepper leaf area as a function of soil type and irrigation water salinity at each growth analysis time. Mossoro-RN, 2010

	EVALUATION PERIODS				
Soils	12 DAE	19 DAE	26 DAE	33DAE	40DAE
PVAe	62,86 a	240,9 a	334,4 a	312b	440,7 b
CXve	38,57 a	180,1 b	248,1 b	337,1 a	553,3 a
ECa	12 DAE	19 DAE	26 DAE	33DAE	40DAE
0,59	79,01 A	261,58 A	402,18 A	403,48 A	613,76 A
2,75	53,75 A	198,75 B	250,86 B	346,63 B	467,49 B
5,00	39,26 A	171,17 B	220,71 B	280,50 C	409,72 C

Values followed by the same letter in the column do not differ by Tukey's test at 5% probability.

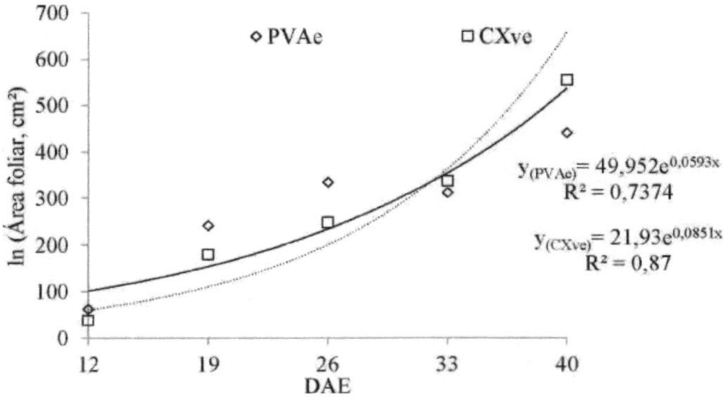

Figura 24. Leaf area of chili peppers grown in PVAe and CXve at different evaluation times. Mossoró-RN, 2010

By unfolding the salinity levels within each evaluation time, there was no significant difference at 12 DAE, at 19 and 26 DAE at 0.59 dS m^{-1} there were higher average leaf area values, while 2.75 and 5.00 dS m^{-1} did not differ. At 33 and 40 DAE, there was a reduction in PA values as salinity increased (Table 22). Leaf area showed similar behavior at all salinity levels, with exponential growth, reaching maximum values at 40 DAE (Figure 25).

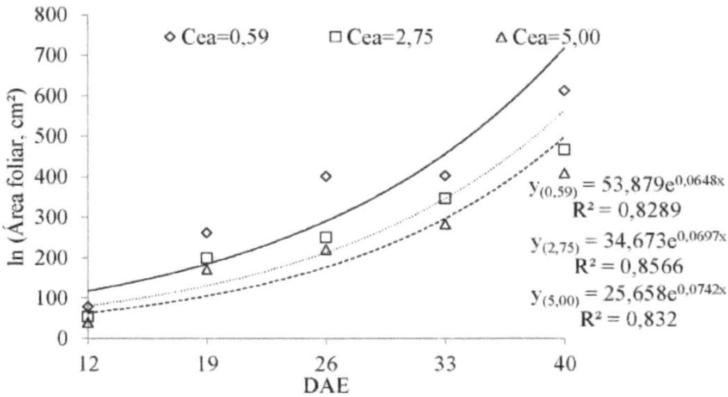

Figura 25. Pepper leaf area under saline stress at different evaluation times. Mossoró-RN, 2010

5.6 MINERAL ELEMENT CONTENT IN THE LEAVES

5.6.1 Honey cultivation

As for the levels of mineral elements in the leaves of the melon tree, there was a significant effect of the soil x water interaction for the levels of K, Na, N, Ca, Mg and Zn. For P, Cu, Mn and Fe, there were isolated effects for soil type and irrigation water salinity levels (Table 24).

Table 24. F-test values and statistical significance for mineral element content in melon leaves under salt stress in two types of soil. Mossoró, 2012

FV	GL	P	K	In	N	Ca	Mg	Cu	Mn	Fe	Zn
Soil	1	$17{,}60^{**}$	$23{,}30^{**}$	$658{,}6^{**}$	$49{,}17^{**}$	$256{,}7^{**}$	$15{,}93^{**}$	$25{,}89^{**}$	$283{,}0^{**}$	$4{,}81^{*}$	3.22^{ns}
Water	2	$75{,}9^{**}$	$19{,}73^{**}$	$71{,}01^{**}$	$0{,}15^{*}$	$41{,}74^{**}$	2.77^{ns}	$9{,}20^{**}$	$6{,}02^{*}$	$54{,}56^{**}$	$36{,}74^{**}$
S X A	2	0.628^{ns}	$4{,}34^{*}$	$8{,}55^{**}$	$15{,}19^{**}$	$17{,}07^{**}$	$5{,}59^{*}$	2.04^{ns}	0.16^{ns}	0.00^{ns}	$6{,}47^{*}$
Res.	12										
CV (%)	-	7,15	7,56	7,16	5,79	6,49	14,13	6,83	10,81	2,66	3,70

**significant at 1% probability,*significant at 5% probability and ^{ns} not significant

As for the K content, at salinity levels of 0.59 and 2.75 dS $_{m-1}$, no differences were observed for the two soils, with the highest values being obtained at a salinity of 5.0 dS m^{-1}. Splitting the salinity levels for each type of soil, it was observed in PVAe and CXve that the highest value was obtained at the highest salinity level and for the others they were statistically similar (Table 25).The results obtained in this work disagree with Sousa (2007) who observed no influence of salinity on K levels.

For Na content, CXVe showed higher values at all salinity levels. In PVAe, the highest value was found at the highest level (5.0 dSm⁻1) and there was no difference in the others. For CXVe, Na levels increased as salinity increased (Table 25). High concentrations of sodium ion can inhibit K absorption by plants through antagonism between these two ions, and potassium ion leakage can occur when Na is replaced by Ca in cell membranes (MARSCHNER, 1995). In fact, some authors have observed the existence of multiple absorption systems with different selectivities for K and Na, which may reflect the plant's need to coordinate the influx of these cations (SCHACHTMAN; LIU, 1999).

The N contents were statistically similar for each salinity level in the two soils studied. For PVAe, the highest N value was obtained at the intermediate level, while for CXVe the highest value was found at the highest level and the others were statistically similar (Table 25). Thus, the average N levels obtained at the different salinity levels are considered adequate for nutrients in melon leaves compared to those suggested by Malavolta *et al*, (1997), who consider a total N content of 30 g kg^{-1}.

Ca levels were higher in CXve at all levels analyzed. In PVAe, the highest content was observed at the lowest salinity level and the others were similar; in CXve, the highest value was found at the highest level, and the others were statistically similar (Table 25). This increase in Ca in the leaves of the melon tree can be explained by the predominance of Ca salts in the irrigation water, especially in plants irrigated with water with a higher salinity level.

For Mg content, the highest values were found for CXVe at all the levels evaluated. In PVAe, the highest level was obtained at the highest salinity level and there was no difference in the others; for CXVe, the values did not differ between salinity levels (Table 25). The increase in Mg levels as a function of salinity can be explained by the concentration effect, because due to the lower dry matter accumulation, the rate of nutrient absorption is faster than its growth.

For Zn, it can be seen that CXVe was higher only at the lowest level and the others were statistically similar. Splitting the salinity levels for each soil type showed that Zn levels increased as a function of salinity for the two soils studied (Table 25). Even though the micronutrient levels are affected by the salinity of the irrigation water, they are still within the range and levels considered adequate for micronutrients, according to the Manual of chemical analysis of soils, plants and fertilizers (EMBRAPA, 2009).

For P content, the highest average was obtained in CXVe (Table 26), and for all salinity levels, there was an increase in this nutrient as a result of the increase in electrical conductivity (Table 25). According to Navarro (2001), the increase in P levels in plants under saline stress may be due to the increased absorption of this nutrient by the root as a result of the stress.

Cu levels were highest in CXVe (Table 26). Analyzing the salinity levels, it can be seen that the levels of this nutrient were lowest at the salinity level of 0.59 dS m^{-1}, and similar in the others (Table 26).

The Mn content in the leaves of the melon tree was highest in CVXe (Table 26). At salinity levels of 2.75 and 5.00 dS m^ 1, they were similar to each other and lower at the less saline level (Table 26), showing the concentration effect of this nutrient.

As for Fe content, the highest values were observed in PVAe (Table 25). However, when analyzing the levels, an increasing behavior was observed as a function of the increase in salts in the irrigation water (Table 26).

The levels of Ca, Mg and Cu are below the range recommended by Embrapa (2009) for melon cultivation.

Plants irrigated with more saline water generally accumulate more macro and micronutrients in the leaves, which allows for greater absorption of water into the plant. However, as the leaf area under saline stress is reduced, production is not increased.

Table 25. Mineral element content in melon leaves during the flowering period as a function of soil type and irrigation water salinity. Mossoró-RN, 2012

Salinity of irrigation water (dS m^{-1})			
	0,59	2,75	5,00
Soils	-	**K (g kg^{-1})**--------------------	
PVAe	38.40Ab	42.72Ab	54.63Aa
CXVe	35.52Aa	37.27Aa	41.43Ba
		-- **Na (g kg^{-1})** ----------------------	
PVAe	3.00Bb	3.28Bb	5.40Ba
CXVe	7.60Ac	10.00Ab	11.93Aa

		-- N (g kg^{-1}) -----------------------	
PVAe	39.14 Ab	44.92 Aa	38.27 Ab
CXVe	34.54 Aab	30.04 Ab	36.39 Aa
		-- Ca(g kg($^{-1)}$)----------------------	
PVAe	10.43Ba	8.35Bb	8.38Bb
CXVe	17.31Aa	11.25Ab	16.34Aa
		-- Mg (g kg^{-1})----------------------	
PVAe	1.92Bb	2.41Aab	3.18Aa
CXVe	3.54Aa	3.05Aa	3.23Aa
------------------	--------------------------	Zn (mg kg^{-1})------------------------------	
PVAe	27.09Bc	32.35Ab	35.15Aa
CXVe	30.96Ab	32.07Aab	34.60Aa

Averages followed by the same letter, upper case in the columns and lower case in the rows, do not differ by Tukey's test at 5% probability.

Table 26. P, Cu, Mn and Fe content in PVAe and CXve as a function of salinity levels. Mossoró, 2012

Soils	P (g $_{kg^-}$^{1})	Cu (mg kg^{-1})	Mn (mg kg^{-1})	Fe (mg kg^{-1})
PVAe	3,02 B	8,09 B	64,95 B	283,03 A
CXve	3,48 A	9,53 A	162,38 A	275,34 B
Salinity levels (dS $_{m-1}$)	P (g kg^{-1})	Cu (mg kg^{-1})	Mn(mg kg^{-1})	Fe(mg kg^{-1})
0,59	2,47 C	8,00 B	100,27 B	256,33 C
2,50	3,17 B	8,97 A	116,23 A	280,09 B
5,00	4,12 A	9,47 A	124,48 A	301,14 A

Means followed by the same letter in the columns do not differ by Tukey's test at 5% probability.

5.6.2 Bean cultivation

In the bean crop, there was a significant effect of the interaction between the soils and the different salinity levels for P, K, Na, N, Ca, Cu, Mn and Zn. Fe levels were only significant for the soils (Table 27).

Table 27. F-test values and statistical significance for mineral element content in bean leaves under salt stress in two types of soil. Mossoró, 2012

FV	GL	P	K	In	N	Ca	Mg	Cu	Mn	Fe	Zn
Soil	1	54,35**	4.32^{ns}	158,9**	2.453^{ns}	15,60**	158,4**	29,88**	467,7**	6,91*	15,54**
Water	2	36,42**	9,25**	145,4**	7,62*	5,42**	135,8**	1.931^{ns}	32,01**	1.66^{ns}	17,59**
S X A	2	9,77**	5,42*	31,91**	4,83**	10,81**	85,45**	7,18**	31,42**	0.11^{ns}	85,18**
Res.	12										
CV (%)	-	8,455	7,10	5,89	6,48	7,88	5,03	6,97	5,35	12,11	2,28

**significant at 1% probability,*significant at 5% probability and ^{ns} not significant

P levels were statistically similar at 2.75 and 5.00 dS m^{-1}, and lowest at 0.59 dS m^{-1} for PVAe, while in CXve they were highest at the highest salt level and similar at 0.59 and 2.75 dS m^{-}$_{1}$. Breaking down the soils within each level, it can be seen that in PVAe the behavior was similar at the 0.59 and 5.00 dS m^{-1} levels, where the highest values of this element were observed in CXve and at the 2.75 dS m^{-1} level there was no statistical difference between the soils (Table 28). Similar results were observed by several authors (LEVITT, 1980; MAAS, GRIEVE, 1987 and LACERDA, 1995). These results highlight the need for caution when carrying out phosphate fertilization in saline environments, as the optimum level of this nutrient for plant production in the absence of salts can be toxic for some plants grown in saline environments (NIEMAM and CLARK, 1976; GRATTAN, MAAS, 1984).

For K, there was no statistical difference for the salinity levels in PVAe, while in CXve the highest value was observed at the lowest salinity level and in the others there was no statistical difference. At salinity levels of 0.59 and 2.75 dS m^{-1}, the K contents were similar in the two soils, while at the highest salinity level the K value was higher in PVAe than in CXve (Table 28). The results found in this study for PVAe corroborate those of Sousa (2006), who found no influence of salinity on K levels. However, Costa *et al.* (2003) and Silva *et al.* (2003) observed an increase in K under conditions of saline stress in string beans, as was found for CXve. These results suggest that the increase in this nutrient under conditions of salinity may be related to the type of soil, since the effects of salinity are more evident in clayier soils.

Na levels were similar at 0.59 and 2.75 dS m^{-1} and higher at 5.0 dS m^{-1} in the two soils studied. At all salinity levels, the highest Na values were observed in CXve, due to the greater accumulation of salts (Table 28).

The high sodium values in the more saline treatments can be explained by the ionic composition of the irrigation water.

As for N, in PVAe the lowest value was observed at 0.59 dS m^{-1} and the other levels were statistically similar, the opposite behavior to that obtained for CXve where the lowest level showed higher N levels and those at 0.59 and 2.75 dS $_{m-1}$ were similar to each other. At the 0.59 and 2.75 dS $_{m-1}$ levels there was no statistical difference, however, at the highest saline level PVAe had higher N values than CXve (Table 28).

The highest Ca values in PVAe were observed at 2.75 and 5.00 dS $m^{-}{}_{1}$, which did not differ statistically from the least saline level, and in CXve, the lowest value was observed at the most saline level, and the others did not differ statistically. For the lowest salinity level, the lowest Ca value was obtained in PVAe, and there was no statistical difference in the other levels (Table 28). For PVAe, which has little Ca in its composition, the addition of this element by water, at not very high values, increases the Ca values in the leaf tissues. In their work, Garcia *et al.* (2007), in greenhouse conditions, and Sousa *et al.* (2010) and Neves *et al.* (2009) in field conditions, found an increase in the concentration of Ca in the aerial part of maize and string bean plants, respectively, grown in saline environments.

In the PVA, the Mg content increased as the EC of the irrigation water increased and in the CXVe the lowest value was observed at the least saline level, although it did not differ statistically from the 2.75 dS m^{-1} level. At 0.59 dS m^{-1}, there was no statistical difference between the two soils; however, at the 0.59 and 2.75 dS m^{-1} levels, the highest Mg contents were observed in PVAe (Table 28).

The Cu levels in the PVAe were lower at the lowest salinity level and similar in the others, while in the CXve there was no statistical difference between the salinity levels. For the 0.59 and 2.75 dS m^{-1} levels, CXve showed higher values than PVAe, and at the highest salinity level, there was no difference between the two types of soil (Table 28).

As for the Mn content, there was no statistical difference for PVAe and in CXve this nutrient increased as the EC of the irrigation water increased. Splitting the salinity levels for each soil, it can be seen that PVAe was superior to CXve in all concentrations (Table 28).

With regard to Zn, it can be seen that this nutrient increased as a result of the increase in salinity for PVAe and, in CXve, the highest value was obtained at 0.59 dS m^{-1} and the lowest at 2.75 dS m^{-1}. For the least saline level, the highest value was observed in CXve and in the others, the highest values of this nutrient were observed in PVAe (Table 28).

Fe levels were higher in PVAe (Table 29), which is explained by the greater amount of this element in PVAe compared to CXve (Table 3).

K, N, Ca and Mg showed values in the leaf tissue below those considered ideal for growing beans (EMBRAPA, 2009).

Table 28. Mineral element content in bean leaves during the flowering period as a function of soil type and irrigation water salinity. Mossoró-RN, 2012

Salinity of irrigation water (dS m^{-1})			
	0,59	2,75	5,00
Soils		**P (g kg^{-1})** -----------------	
PVAe	2.30Bb	4.28 Aa	3.9 Ba
CXVe	3.98 Ab	4.46 Ab	5.65 Aa
		-- **K (g kg^{-1})**-----------------------	
PVAe	12 Aa	11.43Aa	11.63Aa
CXVe	12.84Aa	10.42Ab	9.45Bb
		-- **Na (g kg^{-1})** --------------------	
PVAe	0.56 Bb	0.62 Bb	0.80Ba
CXVe	0.70Ab	0.79 Ab	1.32 Aa
		-- **N (g kg^{-1})**-----------------------	
PVAe	27.92 Ab	25.75 Aa	28.33 Aa
CXVe	29.70 Aa	24.38 Ab	24.00Bb
		-- **Ca (g kg^{-1})** --------------------	
PVAe	9.72 Bb	13.17Aa	11.78 Aab
CXVe	14.56 Aa	13.88 Aa	11.71 Ab
		-- **Mg (g kg^{-1})**---------------------	
PVAe	1.31 Ac	2.25Ab	2.90Aa
CXVe	1.45 Ab	1.72 Bab	1.61Aa
		-- **Cu (mg kg^{-1})**-------------------	
PVAe	9.08Bb	11.03Ba	11.66Aa
CXVe	13.13Aa	12.66Aa	12.25Aa
		-- **Mn (mg kg^{-1})** ------------------	
PVAe	138.98 Aa	133.3 Aa	139.20 Aa
CXVe	52.20 Bc	78.27 Bb	104.77 Ba
		---**Zn (mg kg^{-1})**-------------------	
PVAe	36.95 Bc	41.98Ab	44.93Aa
CXVe	43.25 Ab	36.12Bc	39.37 Ba

Averages followed by the same letter, upper case in the column and lower case in the row, do not differ by the Tukey test at 5% probability.

Table 29. Foliar Fe content in PVAe and CXve bean plants during the flowering period. Mossoró-RN, 2012

Soils	**Fe (mg kg^{-1})**
PVAe	227,55 A
CXve	195,77 B

Averages followed by the same letter in the columns do not differ by the Tukey test at 5% probability.

5.6.3 Maize cultivation

For the levels of mineral elements in maize leaves, there was a significant effect of the combination of soils and EC of the irrigation water for K, Cu, Mn, Fe and Zn. For P and N there was a soil effect, for Na there was an isolated effect of soils and salinity levels, and for Mg it was influenced only by the salinity of the irrigation water (Table 30).

Table 30. F-test values and statistical significance for mineral element content in maize leaves under salt stress in two types of soil. Mossoró, 2012

FV	GL	P	K	In^{+}	N	Ca	Mg	Cu	Mn	Fe	Zn
Soil	1	120,7**	25,30**	7,60*	35,67**	0.29^{ns}	0.739^{ns}	1,67,0**	205,2**	26,10**	6,55*
Water	2	1.00^{ns}	35,48**	11,91**	1.042^{ns}	0.11^{ns}	7,10**	61,67**	101,2**	9,88**	29,85**
S X A	2	0,04*	13,22**	0.49^{ns}	0.206^{ns}	4,11*	1.55^{ns}	22,10**	32,2 **	7,74**	11,82**
Res.	12										
CV (%)	-	7,06	5,51	10,70	6,81	11,01	5,03	5,63	4,54	5,41	6,11

**significant at 1% probability,*significant at 5% probability and ^{ns} not significant

There was no statistical difference in the K content for the PVAe and in the CXve it was lower at the lower salinity level and statistically similar in the others. At the 0.59 dS m^{-1} level, there were no differences between the soils used and at the 2.75 and 5.00 dS $_{m-1}$ levels there were higher values in CXve (Table 31). The higher K values in CXve are due to the greater amount of this element in this type of soil compared to PVAe (Table 3). Higher K concentrations as a function of increasing salinity was also observed by Silva Jûnior *et al.* (2012). The increase in K concentration occurs because absorption in the initial growth period is high, resulting in greater accumulation in plant tissues.

The Cu levels in PVAe were higher at 2.75 and 5.00 dS m^{-1}, but did not differ statistically from the value obtained at 0.59 dS m^{-1}, in CXve the lowest value was observed at the lowest salinity level and the other levels were similar. At the lowest salinity level, there was no statistical difference, but at the other levels, the values for this micronutrient were higher in CXve (Table 31).

Mn decreased as salinity increased in PVAe, while in CXve the lowest value was observed at the lowest salinity level and the other levels were similar. At the 0.59 and 2.75 dS m^{-1} levels, CXve showed higher values, and at the highest level (5 dS m^{-1}) they were similar in both soils (Table 31). The higher Mn values in CXve are a result of the higher concentration of this element compared to PVAe.

The Fe values in PVAe were higher at the highest salinity levels and similar at the others, while CXve showed statistical similarity at all three salinity levels (Table 30). For salinities of 0.59 and 2.75 dS m^{-1}, there was no statistical difference between the soils and at the highest salinity, the highest value was observed in PVAe (Table 31).

Higher Fe values in PVAe may be the result of the greater amount of this element in PVAe (Table 3). The low concentration of Fe in this type of soil can be explained by the antagonism of this micronutrient with excess calcium ion, since CXve has a greater capacity to retain water and nutrients than PVAe.

Zn contents were similar at 0.59 and 2.75 dS m-1 and lowest at the highest salinity level in PVAe, the opposite behavior to that observed for CXve, where the highest value was observed at the highest salinity level and statistically similar at the others (Table 31). At salinity levels of 0.59 and 5.00 dS m^{-1}, the values obtained in PVAe were higher, however, at the highest salinity level, the highest values were observed in CXve.

For P and N, the highest values were observed in PVAe, while CXve showed the highest Na values (Table 32). The greater availability of P in PVAe can be explained by its texture, since the higher the clay content in the soil, the greater the adsorption of P and the lower its availability. Na values were higher in CXve (Table 32). This is due to the greater capacity for water retention and nutrients, so that ion is not leached into the soil.

Na values were highest at the highest salinity level and were similar at the others. The highest Mg values were observed at 5.00 dS m^{-1}, which did not differ from those observed at 2.75 ds m^{-1} and were lowest for water of 0.59 dS m^{-1} (Table 33).

The values of the mineral elements, with the exception of Na, are within the range considered ideal for corn by Embrapa (2009).

Table 31. Mineral element content in maize leaves during the flowering period as a function of soil type and irrigation water salinity. Mossoró-RN, 2012

Salinity of irrigation water (dS m^{-1})			
	0,59	2,75	5,00
Soils		----- **K ($g\ kg^{-1}$)** --------------------	
PVAe	26.9 Aa	28.81Ba	30.25 Ba
CXVe	25.11 Ab	35.53Ab	37.34 Aa
		-Ca ($g\ kg^{-1}$) ----------------------	
PVAe	5.79 Aa	5.03Aa	4.92 Aa
CXVe	4.86 Aa	5.45 Aa	5.88Aa
		-Cu ($g\ kg^{-1}$) ----------------------	
PVAe	10.28 Ab	11,28 Bab	12.42 Bab
CXVe	11.57Ab	17.5Aa	18.98Aa
		-Mn ($mg\ kg^{-1}$) ------------------	
PVAe	45.53 Bc	55.65 Bb	85.35 Aa

CXVe	75.38 Ab	88.53 Aa	90.12 Aa
		-Fe (mg kg^{-1}) ---------------------	
PVAe	130.12 Ab	138.3 Ab	164.03 Aa
CXVe	125.15 Aa	126.83 Aa	127.57 Ba
		-Zn (mg kg^{-1}) ---------------------	
PVAe	34.47 Ab	44.55Aa	38.23 Bb
CXVe	29.07 Bc	37.03 Bb	42.82 Aa

Averages followed by the same letter, upper case in the column and lower case in the row, do not differ by the Tukey test at 5% probability.

Table 32. P, Na and N content in corn leaves during the flowering period as a function of soil type. Mossoró-RN, 2012

Soils	P (g kg^{-1})	Na(g kg($^{-1)}$)	N (g kg^{-1})
PVAe	3,06 A	0,65 B	23,39 A
CXve	2,12 B	0,72 A	19,29 B

Averages followed by the same capital letter in the column do not differ by Tukey's test at 5% probability.

Table 33. Na and Mg content in maize leaves during the flowering period as a function of irrigation water salinity. Mossoró-RN, 2012

Salinity levels (dS m^{-1})	Na(g kg($^{-1)}$)	Mg (g kg^{-1})
0,59 0,62	B	1,39 B
2,50 0,65	B	1.60 AB
5,00 0,81	A	1,70 A

Averages followed by the same capital letter in the columns do not differ by Tukey's test at 5% probability.

5.3.4 Sorghum cultivation

In the sorghum crop, there was a significant interaction between the soil and water factors for Na, Mg, Cu, Mn, Fe and Zn; an isolated effect of the soil and water factors for P, K and Ca; and a soil effect only for N (Table 34).

Table 34. F-test values and statistical significance for mineral element content in sorghum crop leaves under salt stress in two types of soil. Mossoró-RN, 2012

FV	GL	P	K	In	N	Ca	Mg	Cu	Mn	Fe	Zn
Soil	1	0.00^{ns}	$25{,}_{23**}$	$54{,}_{31**}$	$18{,}_{74**}$	0.02^{ns}	$2{,}_{31**}$	$30{,}_{08**}$	$691{,}_{8**}$	0.03^{ns}	0.75^{ns}
Water	2	$18{,}09^{**}$	$4{,}55^{**}$	$35{,}06^{**}$	2.15^{ns}	$11{,}72^{**}$	$51{,}65^{**}$	$10{,}12^{**}$	$132{,}3^{**}$	0.79^{ns}	$454{,}0^{**}$
S X A	2	1.56^{ns}	0.77^{ns}	$40{,}90^{**}$	2.56^{ns}	1.49^{ns}	$6{,}42^{**}$	$7{,}43^{**}$	$11{,}09^{**}$	$6{,}20^{**}$	$15{,}00^{**}$
Res.	12										
CV (%)	-	8,78	8,10	19,18	4,82	3,55	8,02	8,02	3,83	10,97	2,05

**significant at 1% probability,*significant at 5% probability and $_{ns}$ not significant

The Na levels in the PVAe were higher at the second salinity level and statistically similar at the other levels. In the CXve, the levels of this ion increased as the salinity of the irrigation water increased. The 0.59 and 2.75 dS $_{m-1}$ levels showed similar values and the highest concentration of this ion in the CXve at the most saline level (Table 35). The greater accumulation of salts in CXve as a result of the increased salinity of the irrigation water can be explained by the low efficiency of the leaching sheet in this type of soil.

For Mg, the highest values were observed in the PVAe at the 0.59 and 5.00 dS m^{-1} levels, and the lowest value at 2.75 dS m^{-1}, in the CXve no statistical difference was observed for the salinity levels. At the highest and lowest salinity levels, there was no statistical difference between the soils; however, at the intermediate level, CXve showed the highest value (Table 35).

There was no statistical difference in Cu levels in PVAe, while in CXve, the lowest Cu value was observed at the lowest salinity level and the others were similar. At 0.59 dS m^{-1}, there was no significant difference between the soils, and at the other levels, the highest values were observed in CXve (Table 35).

The Mn content in PVAe followed the trend of increasing EC of the irrigation water , being higher at the more saline level. However, in CXve, it was similar at the first and second salinity levels, being higher at the more saline level. For all salinity levels, the highest values of this ion were observed in CXve (Table 35). The higher Mn values in CXve were due to the greater amount of this element in this type of soil (Table 3).

The Fe values did not differ for the salinity levels in the two soils studied. At 0.59 dS $_{m-1}$, the highest value was observed in PVAe; however, at 2.75 dS $_{m-1}$ the opposite behavior occurred, with a higher value in CXve and at the most saline level, no statistical difference was observed between the soils (Table 35). This behavior observed for Fe can be explained by antagonism with Ca ions.

The Zn content increased as the EC of the irrigation water increased, but for CXve it was lower at the most saline level and similar at the others. At the least saline level, there was no difference between the soils; at 2.75 dS m^{-1}, the highest value was observed in CXve, the opposite behavior to that observed at the most saline level, where the highest value was observed in PVAe (Table 35).

For phosphorus, the highest content was observed at the highest salt level and the other levels were statistically similar. These results agree with Lacerda *et al.* (2006), Sousa *et al.* (2007), Farias *et al.* (2009), in fodder sorghum, cowpea and gliricidia plants irrigated with saline water, respectively. However, reductions in leaf phosphorus levels under saline stress were observed by Ferreira *et al.* (2007).

P was higher in CXve and N in PVAe. Considering the salinity levels, the highest potassium content was found at the 5.00dS m^{-1} level, although it did not differ statistically from the intermediate level, and the lowest value was found at 0.59 dS m^{-1} (Table 36). The P and N values observed are below those recommended for sorghum by Embrapa (2009).

Table 35. F-test values and statistical significance for mineral element content in leaves of the sorghum crop under salt stress in two types of soil. Mossoró, 2012

Irrigation water salinity (dS m^{-1})			
Soils	0,59	2,75	5,00
		---Na (g kg^{-1}) ------------------	
PVAe	1.30 Ab	2.85 Aa	1.20Bb
CXVe	1.21 Ac	3.46 Ab	6.31 Aa
		---Mg (g kg^{-1})-----------------	
PVAe	1.07 Ab	1.47 Ba	1.70 Aa
CXVe	0.86 Ab	1.88 Aa	1.82 Aa
		---Cu (mg kg^{-1})---------------	
PVAe	10.62 Aa	12.35 Ba	10.65Ba
CXVe	11.38 Ab	14.28 Aa	15.73Aa
		---Mn (mg kg^{-1})---------------	
PVAe	45.53 Bc	51.48 Bb	70.61 Ba
CXVe	75.38 Ab	84.6 Ab	96.63 Aa
		---Fe (mg kg^{-1}) ----------------	
PVAe	163.4 Aa	130.15 Ba	134.98 Aa
CXVe	129.18 Ba	159.17 Aa	136.55 Aa
		---Zn (mg kg^{-1})---------------	
PVAe	31.80Ac	41.05 Bb	47.80 Aa
CXVe	33.03 Ab	43.45 Aa	45.18 Ba

Averages followed by the same letter, upper case in the column and lower case in the row, do not differ by the Tukey test at 5% probability.

Table 36. K and N content in sorghum leaves during the flowering period in PVAe and CXCve and P and K as a function of irrigation water salinity. Mossoró-RN, 2012

Soils	**K (g kg^{-1})**	**N (g kg^{-1})**
PVAe	33,04 B	22,11 A
CXve	40,05 A	20,04 B
Salinity levels (dS m^{-1})	**K (g kg^{-1})**	**P (g kg^{-1})**
0,59	33,85 B	1,82 B
2,50	36.80 AB	2,32 B
5,00	38,99 A	2,47 A

Averages followed by the same letter, capitalized in the columns, do not differ by the Tukey test at 5% probability.

5.3.5 Pepper cultivation

For chili peppers, there was a significant interaction for P, K, Na, N, Mg, Cu and Mn. Soil and salinity had an isolated effect on Ca and Fe levels, while Zn only showed a soil effect (Table 37).

Table 37. F-test values and statistical significance for mineral element content in pepper crop leaves under salt stress in two types of soil. Mossoró, 2012

FV	GL	P	K	In$^+$	N	Ca	Mg	Cu	Mn	Fe	Zn
Soil	1	6,43*	48,54**	45,36**	62,45**	19,71**	73,87**	224,4**	2113,7**	0.006^{ns}	73,902**
Water	2	32,94**	4,93**	79,37**	19,68**	76,94**	60,07**	19,96**	36,51**	5,641*	2.607^{ns}
S X A	2	7,14**	34,19**	19,61**	9,30**	0.15^{ns}	6,22*	120,92**	11,52**	1.621^{ns}	0.233^{ns}
Res.	12										
CV (%)	-	11,76	1,43	11,04	6,43	3,97	7,55	6,95	4,33	14,92	14,42

**significant at 1% probability,*significant at 5% probability and ^{ns} not significant

The P levels at 2.75 and 5 dS m^{-1} were similar and did not differ from 0.59 ds m^{-1} for PVAe, and in CXve, the values of this element increased as the EC of the irrigation water increased. At the first and second levels of salinity, there was no difference between the soils, and at the most saline level there was a higher concentration of P in CXve (Table 38). The P concentrations in the leaves were lower than those considered ideal for the crop by Embrapa (1999) and those found by Villas Bôas (2001), on average 3.0 g kg^{-1}. However, the low P values are not the result of salinity, since the levels of this element increased with saline stress in the two soils studied (Table 38), which may have occurred due to a probable deficiency of this element in the soils.

K in PVAe increased as the salinity of the irrigation water increased, however, in CXve the highest value was observed at the lowest salinity level and the others were statistically similar. At levels of 0.59 and 2.75 dS m^{-1}, the highest values were obtained at CXve and at the highest salinity level no statistical difference was observed (Table 38). Overall, the K levels were above the appropriate range for peppers (35.0-45.0 g kg^{-1}), according to Jones Júnior *et al.* (1991), which demonstrates the concentration effect of this nutrient in the flowering phase.

Na levels increased as the EC of the irrigation water increased in the PVAe, while in the CXve the lowest value was found at the least saline level and the others were similar. At levels of 0.59 and 2.75 dS m^{-1}, there was no statistical difference between the soils, while at 5.00 dS m^{-1}, the highest values of this ion were observed in PVAe (Table 38). The increase in Na levels as the ECa increased is explained by the abundance of this element in irrigation water.

N was lower at the second salinity level and at the extremes was statistically similar in PVAe, while in CXve there was no statistical difference between salinity levels. For the highest and lowest salinity of the irrigation water, CXve showed lower values, while at the intermediate level the two soils were similar (Table 38). The high concentration of N in the leaves in response to treatments under higher soil EC can be explained by the accumulation of nitrate in the vacuoles, according to Martinez and Cerda (1989). According to these authors, the accumulation of nitrate in the leaves can influence the cation-anion balance and the osmotic balance. The N values obtained in this work were higher than those obtained by Villas Bôas (2001), from 36 to 38 g kg^{-1}.

Mg levels were lower at the lowest salinity level and at the other levels they were similar for the two soils studied. At the lowest EC of the irrigation water, there was no difference in Mg content between the soils, and at the other levels, the highest values were observed in PVAe (Table 38). It can be seen that even under saline stress, foliar Mg levels were within the appropriate range for peppers (3.0-10.0 g kg^{-1}), according to Jones Júnior *et al*. (1991).

Cu values decreased with increasing salinity in PVAe, and in CXve, the highest value was observed at the most saline level and the others were similar. For the 0.59 and 2.75 dS m^{-1} levels, PVAe showed higher Cu values, while at the most saline level there was a higher concentration of this element in CXve. At all salinity levels, CXve had higher Mn values than PVAe (Table 38), showing the concentration effect of this element up to the flowering stage. In all the treatments, the Cu levels were between the values considered adequate for the crop (EMBRAPA, 2009).

The Mn levels were highest at 5.00 dS m^{-1}, even though they did not differ statistically from 2.75 dS m^{-1}, which was similar to the lowest level for PVAe; however, in CXve there was an accumulation of this element as a result of the increase in salinity of the irrigation water (Table 39). The Mn levels observed in this experiment are in line with the range considered adequate by Embrapa (2009).

Ca and Zn values were higher in PVAe. The Ca content was highest at the highest saline level and similar at the other levels, with values within the range recommended by Embrapa (2009). Fe was highest at 0.59 dS m^{-1}, which was statistically similar to the intermediate level, and this did not differ from the value obtained at the most saline level (Table 39). At the levels of 0.59 and 5.00dS m^{-1}, the Fe values were above the appropriate range for pepper crops (EMBRAPA, 2009).

Table 38. Mineral element content in the leaves of the pepper crop during the flowering period as a function of soil type and irrigation water salinity. Mossoró-RN, 2012

Salinity of irrigation water (dS m^{-1})			
Soils		**P (g kg^{-1})**	
	0,59	2,75	5,00
PVAe	1.09 Ab	1.37 Aab	1.51 Bab
CXVe	0.93 Ac	1.58 Ab	2.07 Aa
		----K (g kg^{-1}) ----	
PVAe	61.62 Bc	65.19 Bb	67.71 Aa
CXVe	69.77 Aa	67.11 Ab	67.02 Ab
		---- **Na (g kg^{-1})**----------------	
PVAe	1.71 Ac	2.72 Ab	4.7 Aa
CXVe	1.43 Ab	2.38 Aa	2.6 Ba
		-- N (g kg^{-1}) ---	
PVAe	74.03 Aa	52.44 Ab	69.53Aa
CXVe	54.13 Ba	49.23 Aa	50.69 Ba
		---- **Mg (g kg^{-1})** ----	
PVAe	4.7 Ab	7.61 Aa	8.16 Aa
CXVe	3.94 Ab	5.23 Ba	5.86 Ba
		--- **Cu (mg kg^{-1})** ---	
PVAe	22.32 Aa	15.23 Ab	11.85 Bc
CXVe	7.8 Bb	8.32 Bb	13.82 Aa
		---- **Mn (mg kg^{-1})** ----	
PVAe	42.75 Bb	46.52 Bab	51.13 Bab
CXVe	113.88 Ac	131.55 Ab	143.02 Aa

Averages followed by the same letter, upper case in the column and lower case in the row, do not differ by the Tukey test at 5% probability.

Table 39. Ca and Zn content in the leaves of the pepper crop during the flowering period in PVAe and CXve and Ca and Fe as a function of irrigation water salinity. Mossoró-RN, 2012

Soils	**Ca(g kg$^{(-1)}$)**	**Zn (mg kg^{-1})**
PVAe	12,30 A	73,58[a]

CXve	11,32 B	40,31 B
Salinity levels (dS m^{-1})	**Ca (g kg^{-1})**	**Fe (mg kg^{-1})**
0,59	10,52 B	330,14 A
2,50	11,21 B	389.88 AB
5,00	13,71 A	246,57 B

Averages followed by the same capital letter in the columns do not differ by the Tukey test at 5% probability.

6. CONCLUSION

EC values increased with increasing salinity at 15 and 36 DAE, with this effect being greater in clay soils.

There were reductions in the total dry matter of the aerial part (MSTPA) of the melon tree as the salinity of the irrigation water increased for RQ, PVAe, CXve and RY. The greatest reductions in leaf area were seen in VX, RY and CXve. At 36 DAE, the highest PA values were observed in RQ and the lowest in VX.

The highest peaks in stomatal conductance occurred at 8:00 a.m., with a more significant reduction until 12:00 p.m., continuing to decrease until 4:00 p.m., but with less intensity. Higher conductance values were observed in the RY and VX when water with lower salinity was irrigated.

There were reductions in MSTPA for the bean, maize and sorghum crops as salinity increased, with the greatest decreases observed for the bean and sorghum crops at CXve.

The melon crop showed a reduction in PA with increasing salinity at 33 DAE and for beans, maize, sorghum and peppers at 33 and 40 DAE. The greatest reductions in PA occurred in PVAe for beans, maize and sorghum; in CXve for peppers.

There was an interaction between soil types and salinity levels for mineral elements in all crops.

7. BIBLIOGRAPHICAL REFERENCES

ALVES, G. S.; SANTOS, D; SILVA, J. A.; NASCIMENTO, J. A. M; CAVALCANTE, L. F.; DANTAS, T. A. G. Nutritional status of pepper grown in soil treated with different types of biofertilizers. **Revista Acta Scientiarum**, Maringà, v. 31, p.661-665, 2009.

ASSIS JÛNIOR, J. O. et al. String bean productivity and soil salt accumulation as a function of leaching fraction and irrigation water salinity. **Engenharia Agricola**, Jaboticabal, SP, v.27, p.702-713, 2007.

AYERS, R. S.; WESTCOT, D. W. **Water quality in agriculture**. Translated by GHEYI, H. R.; MEDEIROS, J. F; DAMASCENO, F. A. V. 2.ed. Campina Grande: UFPB, 1999. 153p. FAO Irrigation and Drainage Studies, 29.

AYRES, IAN. "Fair Driving: Gender and Race Discrimination in Retail Car Negotiations." Harvard Law Review, February 1991, 104(4), pp. 817-72.

AZEVEDO NETO, A.D.; TABOSA, J.N. Salt stress in maize seedlings: Part II. Distribution of cationic macronutrients and their relationship with sodium.

Revista Brasileira de Engenharia Agricola e Ambiental, Campina Grande, v.4, n.2, p.165-171, 2000.

BARROS, A. D. **Management of drip irrigation with different levels of water salinity in the melon crop.** Botucatu, 2002. 124 f. Thesis (Doctorate in agronomy). Faculty of Agronomic Sciences, São Paulo State University, 2002.

BATAGLIA, O.C.; DECHEN, A.R. and SANTOS, W.R. Principle of foliar diagnosis. In: ALVAREZ V., V.H.; FONTES, L.E.F. and FONTES, M.P.F. **Os solos nos grandes dominios morfológicos do Brasil e o desenvolvimento sustentàvel.** Viçosa, MG, SBCS, UFV, DPS, 1996. 930p.

BEZERRA, M. E. J.; LACERDA, C. F.; SOUSA, G. G.; GOMES, V. F. F.; MENDES FILHO, P. F. Biomass, microbial activity and AMF in maize/corn-bean crop rotation using saline waters. **Revista Ciência Agronômica**, Fortaleza: UFC, v.41, p.562-570, 2010.

BLANCO, F.F.; FOLEGATTI, M.V.; GHEYI, H.R.; FERNANDES, P.D. Growth and yield of corn irrigated with saline water. **Scientia Agricola**, Piracicaba, v.65, n.6, p.574-580, 2008.

BOTiA, P. *et al.* Yield and fruit quality of two melon cultivars irrigated with saline water at different stages of development. **European Journal of Agronomy**, v. 23, n. 3, p. 243-253, 2005.

BRAZIL. Ministry of Agriculture. Pedological Research Division, DNPEA. Exploratory survey-recognition of soils in the state of Rio Grande do Norte. Recife, Agreement MA/ DNPES-SUDENE/DRN, MA/USAID/BRASIL. 1971. 530p. (Technical Bulletin, 21).

BRAY, E. A.; BAILEY-SERRES, J.; WERETILNYK, E. Responses to abiotic stresses. In: BUCHANAN, B. B., GRUISSEM, W.; JONES, R. L. (Ed.). Biochemistry and Molecular Biology of Plants. Rockville: **ASPP**, p.1158-1203. 2000.

CARMO FILHO, F. do; OLIVEIRA, O. F. de. **Mossoró: a semi-arid municipality: climatic characterization and floristic aspect.** Mossoró: ESAM, 1989. 62 p. (ColeçaoMossoroense, 672, series B).

CAVALCANTE, L. F.; VIEIRA, M. da SILVA; SANTOS, A. F. dos; OLIVEIRA, W. M.; NASCIMENTO, J. A. M. do. Saline water and liquid bovine manure in the formation of guava seedlings of the paluma cultivar. **Revista Brasileira de Fruticultura**, v.32, n.1, p. 251-261, 2010.

CHARTOZOULAKIS, K.S. Effects of NaCl salinity germination, growth and yield of greenhouse cucumber. **The Journal of Horticultural Science.** p. 115119,1992.

CLOUSE - Available at: http://www.clause-vegseeds.com/uk/home/index.php. Consulted in January 2011.

COSTA, P. H. A.; SILVA, J. V.; BEZERRA, M. A.; ENÉAS FILHO, J.; PRISCO, J. T.; GOMES FILHO, E. Growth and levels of organic and inorganic solutes in cultivars of Vigna unguiculata submitted to salinity. **Revista Brasileira de Botânica**, v.26, p.289-297, 2003.

CRUCIANI, D. E. **Drainage in agriculture.** 4. ed, Sao Paulo: Nobel, 1987, 337p.

CUENCA, R. H. **Irrigation system design: an engineering approach**. Englewood Cliffs: Prentice Hall, 1989. 552p.

DANTAS, D. da C.; MATOS, J. de A. de; MEDEIROS, J. F. de; PORTO FILHO, F. de Q.; MIRANDA, N. de M. Electrical conductivity of the soil saturation extract from 1:2 extracts by the reduced principal axis method. **Revista Irriga**, Botucatu, v.10, n.4, p. 335-340, 2005.

DANTAS, J.P.; MARINHO, F. J. L.; FERREIRA, M.M.M.; AMORIM, M.S.N.; ANDRADE, S.I.O.; SALES, A.L. Evaluation of caupis genotypes under salinity. **Revista Brasileira de Engenharia Agricola e Ambiental**, Campina Grande, v.6, n.3, 2002.

DIAS, N.S.; DUARTE, S.N.; GHEYI, H.R.; MEDEIROS, J.F.; SOARES, T.M. Manejo da fertirrigaçao e controle da salinidade do solo sob ambiente protegido, utilizando extratores de soluçao do solo. **Revista Brasileira de Engenharia Agricola e Ambiental**, Campina Grande, PB. v.9, n.4. 2005.

DIÓGENES, T. B. A. **Response of papaya to doses of potassium, boron, zinc, copper, magnesium and nitrogen sources**. Dissertation (Master's Degree in Soil Science). Federal Rural University of the Semi-Arid Region. Mossoró, 2012.45f.

DOURADO NETO, D.; NIELSEN, D. R.; HOPMANS, J. W.; REICHARDT, K.; BACCHI, O. O. S.; LOPES, P. P. Program for making a soil water retention curve, Van Genuchten model. Soil Water Retention Curve, SWRC (version 3.00 beta). University of Sao Paulo, Piracicaba, SP, Brazil, 2001.

EMBRAPA. **Manual of chemical analysis of soils, plants and fertilizers**. Embrapa Informaçao Tecnològica. 2ª revised and expanded edition. 2009.

EMBRAPA. Brazilian Agricultural Research Corporation. Available at http://www.embrapa.br/imprensa/artigos/2009/qualidade-da-agua-subterranea-e-riscos-para-irrigacao in June 2010. Consulted in June 2012.

ERNESTO SOBRINHO, F. **Caracterizaçâo, Gênese e Interpretaçâo para Uso de**

Solos Derivados de Calcàrio da Regiâo da Chapada do Apodi Rio Grande do Norte. Dissertation (Master's Degree in Soil Science). Federal University of Viçosa. 1980. 133p.

FARIAS, S. G. G. Salt stress on initial growth and mineral nutrition of gliricidia (*Gliricidia sepium* (Jacq.) Kunth ex Steud) in nutrient solution. **Revista Brasileira de Ciência do Solo**, v.33, p. 1499-1505, 2009.

FEIGIN, A. Interactive effects of salinity and ammonium/nitrate ratio on growth and chemical composition of melon plants. **Journal of Plant Nutrition**, New York, v.13, n.10, p.1257-1269, Oct.1990.

FERNANDES, M.B.; CASTRO, B.F.; FERNANDES, J.R.; AQUINO, V.L.B.; ALVES, J.S.; GÓIS, F.C. Adaptability of sorghum cultivars to saline soils in the saline micro-region of the state of Rio Grande do Norte. **Pesquisa agropecuâria Brasileira**, 27 (2): 222, 1992.

FERNANDES, M.B.; CASTRO, J.R.; AQUINO, B.F.; FERNANDES, V.L.B.; ALVES, J.S.; GÓIS, F.C. Evaluation of forage sorghum cultivars in soils affected by salts in the state of Rio Grande do Norte. **Pesquisa agropecuâria Brasileira**, 29 (2): 255,1994.

FERREIRA, P. A.; GARCIA, G. O.; NEVES, J. C. L.; MIRANDA, G. V.; SANTOS, D. B. Relative yield of corn and leaf contents of nitrogen, phosphorus, sulfur and chlorine as a function of soil salinity. **Revista Ciência Agronômica**, v.38, n.1, p.7-16, 2007.

FLOWERS, T. J.; FLOWERS, S. A. Why does salinity pose such a difficult problem for plant breeders. **Agricultural Water Management**, v. 78, n.1, p. 1524, 2005.

FRANCO, J. A.; FERNANDEZ, J. A.; BANÓN, S.; GONZALEZ, A. Relationship between the effects of salinity on seedling leaf area andfruit yield of six muskmelon cultivars. **HortScience**, Alexandria, v.32, n.4, p.642-644, 1997.

FRANÇÔIS, L. E.; DONOVAM, T.; MASS, E. V. Salinity effects on seed yield, grouth and germination of grain sorghum. **Agronomy Journal**, v.76, p.742-744, 1984.

FREIRE FILHO, F. R.; LIMA, J. A. A.; RIBEIRO, V. Q. **Feijâo-caupi: Avanços tecnológicos.** Brasilia: Embrapa Informaçao Tecnològica, 2005. 519p.

FREIRE, M.B.G.S. Water conductivity of soils in Pernambuco in response to electrical conductivity and RAS of irrigation water. **Brazilian Journal of Agricultural and Environmental Engineering**. Campina Grande, PB. v.7, n.1. 2003.

FOLEGATTI, M.V. BLANCO, F.F. Vegetative development of grafted cucumber irrigated with saline water. **Scientia Agricola**, Piracicaba, v.57 n.3, 2000.

GARCIA, G. O.; FERREIRA, P. A.; MIRANDA, G. V.; OLIVEIRA, F. G.; SANTOS, D. B. Physiological indices, growth and yield of maize irrigated with saline water. **Revista Irriga**, v. 12, n. 03, p. 307-325, 2007.

GHEY, H. R.; DIAS, N. S.; LACERDA, C. F. **Management of salinity in agriculture: basic and applied studies.** Fortaleza: INCT sal, 2010. 472p.

GOMIDE, M.B.; LEMOS, O.V.; TOURINO, D.; CARVALHO, M.M.; CARVALHO, J.G.; DUARTE, C.S. Comparison between methods for determining leaf area in Mundo Novo and Catuai coffee trees. **Ciência Pràtica**, Lavras, v.1, n.2, p.118-123, 1977.

GONDIM, F. A.; GOMES FILHO, E.; MARQUES, E. C.; PRISCO, J. T. Effects of H2O2 on growth and solute accumulation in maize plants under salt stress. **Revista Ciência Agronòmica**, v. 42, n. 02, p. 373-381, 2011.

GRATTAN, S. R.; GRIEVE, C. M. Salinity-mineral nutrition relations in horticultural crops. **Scientia Horticulturae**, Lexington, v, 78, p.127-157, 1999.

GRATTAN, S. R.; MAAS, E. V..Interactive effects of salinity and substrate phosphate on soybean.**Agronomy Journal**, v.76, p.668-676, 1984.

GURGEL, M.T.; GHEYI, H.R.; FERNANDES, P.D.; SANTOS, F.J.S.;NOBRE, R.G. Initial growth of guava rootstocks irrigated with saline water. **Revista Caatinga,** Mossoró, RN. v. 20, n. 2. 2007.

HUERTA, S.A. Comparison of laboratory and field methods for measuring coffee tree area. *Cenicafé*, **Chinchina**, v.13, n.1, p.33-42, 1962.

HOLANDA, J. S.; AMORIM, J.R.A.; FERREIRA NETO, M. HOLAMDA, A. C. Water quality for irrigation. In: GHEI, H. R.; DIAS, N. S.; LACERDA, C. F. Management of salinity in agriculture: basic and applied studies. Fortaleza, INCTSal, 2010. 472p.

JONES JÛNIOR. JB; WOLF B; MILLS HA. 1991. Plant analysis handbook: a practical sampling, preparation, analysis and interpretation guide. **Athens: Micro-Macro**, 213p.

LACERDA, C. F. Effects of salinity on the development and mineral composition of cowpea (*Vigna unguiculata* (L.) Walp.) and the use of Ca^{2+} as a means of mitigating these effects. **Dissertation** (Master's Degree in Soils and Plant Nutrition), 87f- Universidade Federal do Cearà, Fortaleza, 1995.

LACERDA, C. F. DE.; MORAIS, M. M. M.; PRISCO, J.T.; GOMES FILHO, E.; BEZERRA, M. A. Interaction between salinity and phosphorus in forage sorghum plants. **RevistaCiência Agronômica**, v. 37, n. 3, p.258-263, 2006.

LACERDA, C. F. Interaction between salinity and mineral nutrition. In: NOGUEIRA, R. J. . C., ARAÙJO, E. L., WILLADINO, L. G., CAVALCANTE, U. (ed.) **Environmental stresses: damage and benefits to plants**. Recife: UFRPE, 2005, p. 127-137.

LAUCHI, A.; EPSTEIN, E. Mechanisms of salt tolerance in plants.**California Agriculture**, v. 38, n.10, p.18-21. 1984.

LEONARDO, M.; BROETTO, F.; VILLAS BÔAS, R. L.; ALMEIDA, R. S.; MARCHESE, J. A. Pepper fruit production in different salt concentrations. **Irriga**, Botucatu, v. 12, n. 1, p. 73-82, 2007.

LIMA, C.F.G. de S.; OLIVEIRA, F. de A.; MEDEIROS, J.F.; OLIVEIRA, M.K.T, OLIVEIRAFILHO, A.F. Mathematical models for estimating the leaf area of caupi beans. **Caatinga**, Mossoró-RN, v.21, n.1, p.120-127, 2008.

LiMA, C. J. de; OLiVEiRA, F. de A. de; MEDEiROS, J. F.; OLiVEiRA, M. K. T.;

ALMEIDA JÙNIOR, A. B. de. Response of caupi beans to irrigation water salinity. **Revista Verde de Agroecologia**. Mossoró-RN. v.2, n.2, pg 79-86, 2007.

LIMA J.R; SILVA, A.L.P. Study of the salinization process to indicate measures to prevent saline soils. **Enciclopédia biosfera**, Goiânia, GO.v.6, n. 11.2010.

MAAS, E. V.; HOFFMAN, G. J. Crop salt tolerance - Current assessment. **Journal of Irrigation and Drainage Division**, v.103, n.Ir^2, p.115134, 1977.

MAAS, E. V.; GRIEVE, C. M..Sodium-induced deficiency in salt-stressed corn.**Plant, Cell and Environment**, v.10, p.559-564, 1987.

MAAS, E.V. Salt tolerance of crops.**Applied Agricultural Research**, New York, v.1, n.1, p.12-26, 1986.

MAIA, C.E.; MORAIS, E.C.M.; OLIVEIRA,M.Classification of the ionic composition of irrigation water using multiple linear regression**. Revista Brasileira de Engenharia Agricola e Ambiental.**Campina Grande, PB. v.5, n.1. 2001.

MAIA, C. E.; MORAIS, E. R. C.; OLIVEIRA, M. Water quality for drip fertigation. I: Chapada do Apodi-RN region. **Engenharia na Agricultura, Viçosa**, v.5, p.281-292, 1997.

MALASH, N.; FLOWERS, T. J.; RAGAB, R. Effect of irrigation systems and water management practices using saline and nom-saline water on tomato production.**Agricultural Water Management**, v.78, p.25-38, 2005.

MALAVOLTA, E.; VITTI, G.C.; OLIVEIRA, S.A. **Avaliaçâo do estado nutricional das plantas: Principios e aplicções.** Piracicaba: Brazilian Association for Potash and Phosphate Research, 1989. 201p.

MALAVOLTA, E. VITTI, G. C.; OLIVEIRA, S. A. **Evaluation of plant nutritional status: principles and applications**.Piracicaba: POTAFOS, 1997. 319 p.

MARSCHNER, H. **Mineral nutrition of higher plants.** 2ed. London: Academy Press, 1995. 889p.

MARTINEZ, V.; CERDA, A. Nitrate reductase activity in tomato and cucumber leaves as influenced by NaCl and N source. **Journal of Plant Nutrition**, New York, v.12, p.1335 - 1350, 1989.

MEDEIROS, J.F. de; LISBOA, R. de A.; OLIVEIRA, M. de. **Characterization of groundwater used for irrigation in the Chapada do Apodi melon-producing area**. Revista Brasileira de Engenharia Agricola e Ambiental, 7 : 469-472. 2003.

MEDEIROS, J. F. de. **Management of saline irrigation water in a greenhouse cultivated with peppers.** Piracicaba, 1998, 152 p. Thesis (Doctorate in Agronomy/Irrigation and Drainage) - Escola Superior de Agricultura "Luiz de Queiroz", Universidade de Sao Paulo, Piracicaba, SP, 1998.

MELONI, D. A.; OLIVA, M. A.; RUIZ, H. A.; MARTINEZ, C. A. Contribution of proline and inorganic solutes to osmotic adjustment in cotton under salt stress. **Journal of Plant Nutrition**, v. 24, n. 3, p. 599-612, 2001.

MEIRI, A.; PLAUT, Z.; PINCAS, L. Salt tolerance of glasshouse grown muskmelon.**Soil Science,** v. 131, p 189-193, 1981.

MEIRELES, A. C. M.; Evaluation of the impact of fertigation on cambisols in Chapada do Apodi, Ceará. **Revista de Ciência Agronòmica**, v. 34, n.2, p. 207212, 2003.

MENDLINGER S; FOSSEN M. Flowering, vegetative growth, yield, and fruit quality in muskmelons under saline conditions.**The Journal of American Society for Horticultural Science.**p.118: 868-872, 1993.

MORAIS, E. R.; MAIA, C. E,; OLIVEIRA, M. Quality of water for irrigation in analytical samples from the database of the department of soils and geology of the higher school of agriculture of Mossoró, Mossoró-RN. **Caatinga**, vol. 11, p 7583, dec, 1998.

MUALEM, Y. A new model for predicting the hydraulic conductivity of unsaturated porous media.**Water Resources Research**.v.12, p.513-522, 1976.

MUNNS, R. Genes and salt tolerance: bringing them together. **New Phytologist**, v. 167, n. 03, p. 645-663, 2005.

MUNNS, R. Comparative physiology of salt and water stress. **Plant, Cell and Environment**. 25,239-250, 2002.

NASCIMENTO, I. B.; MEDEIROS, J. F.; MIRANDA, N. O.; SILVA, M. C. C.; ALVES, L. A.; LISBOA, R. A.; SILVA JÙNIOR, M. J. da. Vegetative development of melon grown under different intensities of soil preparation with and without mulching. In: Brazilian Congress of Agricultural Engineering, 29, 2000. Fortaleza, **Proceedings** . Fortaleza: SBEA, 2000. CDROM.

NASCIMENTO, I.B. *et al*. Estimation of the leaf area of the melon tree. **Horticultura Brasileira**, v.20, n.4, p.555-558,2002.

NAVARRO, J. M.; BOTELLA, M. A.; MARTINEZ, V. Yield and fruit quality of melon plants grown under saline conditions in relation to phosphate and calcium nutrition. **Journal of Horticultural Science & Biotechnology**, v. 74, n. 5, p. 573-578, 1999.

NAVARRO, J. M.. Phosphorus uptake and translocation in salt-stressed melonplants. **Journal of Plant Physiology**, v.158, p.375-381, 2001.

NEVES, A. L. R.; LACERDA, C. F.; GUIMARÂES, F. V. A.; HERNANDEZ, F. F. F.; SILVA, F. B. da.; PRISCO, J. T.; GHEYI, H. R. Biomass accumulation and nutrient extraction by string bean plants irrigated with saline water at different stages of development. **Revista Ciência Rural**, v.39, p.758765, 2009.

NIEMAN, R. H.; CLARK, R. A. Interactive effects of salinity and phosphorus nutrition on the mature photosynthesizing corn leaves. **Plant Physiology**, v.57, p157-161, 1976.

OLIVEIRA, F. A., MEDEIROS, J.F., OLIVEIRA, M.K.T., LIMA, C.J.G.S., ALMEIDA JÙNIOR, A. B., AMÂNCIO, M.G. Initial development of popcorn corn with water of different salinity levels. **Agrària.**v.4, p.149-1555, 2009.

OLIVEIRA, J.A. Influence of the stage of maturity of forage sorghum, Sorghum bicolor (L) Moench, on the yield and quality of forage and respective silages. Master's **dissertation**, 50p, UFC: Fortaleza, 1983.

OLIVEIRA, M.; MAIA, C.E. **Qualidade fisico-quimica da água para irrigaçâo em diferentes aquiferos na área sedimentar do estado do Rio Grande do Norte.** Campina Grande, DEAg/UFPB, v.2, n.1, p.17-21, 1998.

OULD AHMED, B. A.; MORITANI, M. I. S. **Effect of saline water irrigation and manure application on the available water content, soil salinity, and growth of wheat**. Agricultural Water Management, v.97, p.165-170, 2010.

O'Leary, J.W. High humidity overcomes lethal levels of salinity in hydroponically grown salt-sensitive plants. **Plant and Soil**, Dordrecht, v.42, P.717- 721, 1971.

PATEL, P. R. et al. Impact of saline water stress on nutrient uptake and growth of cowpea. **Brazilian Journal of Plant Physiology,** Campos dos Goytacazes, RJ, v.22, p.43-48, 2010.

PEREIRA, F. A. de L. Tolerance of melon cultivars to salinity. **Dissertation** (Master's Degree in Irrigation and Drainage), Universidade Federal Rural do Semi-Àrido - Mossoró, 2011. 81 f.

PIZARRO, F. **Drenaje agricola y recuperacion de suelos salinos**. Madrid: Editora Agricola Espanola, S.A. 1985, 542p.

PIZARRO, F. **Riegos localizados de alta frequencia: goteo, microaspersión, exudación**. 2 ed. Madrid: Mundi Prensa, 1990, 459p.

REBOUÇAS, A.; FILHO, M.; BENOIT H. **Potiguar Basin - Hydrogeological study**. Recife-PE: SUDENE - Documents Division, 1967.

RICHARDS, L.A. **Diagnosis and improvement of saline and alkali soils**. Washington: United States Department of Agriculture, 1954. 160p. Agriculture Handbook, 60.

RHOADES, J. D.; KANDIAH, A.; MASHALI, A. M. **Use of saline water for agricultural production.** Campina Grande: UFPB, 2000. 117p. FAO Irrigation and Drainage Studies, 48.

RHOADES, J. D.; KANDIAH, A.; MASHALI, A. M. **The use of saline water for**

crop production. Rome: FAO, 1992. 133p. (FAO Irrigation and Drainage Paper, 48).

RHOADES, J. D.; MIYAMOTO, S. **Testing soils for salinity and sodicity**. In: WESTERMAN, R.L. (ed.) Soil testing and plant analysis. Madison: Soil Science Society of America, 1990. chap. 12, p.299-336. (Book Series, 3).

SANTANA, M. J. Effects of irrigation water salinity on the sprouting and initial development of sugarcane (*Saccharum* spp) and on soils with different textural levels. **Ciência e Agrotecnologia**, Lavras, v. 31, n. 5, p. 14701476, 2007.

SCHACHTMAN, D.; LIU, W. Molecular pieces to the puzzle of the interaction between potassium and sodium uptake in plants. **Trends Plant Science,** v. 04, n. 07, p. 281-287, 1999.

SHALHEVET, J.; KAMBUROV, J. **Irrigation and salinity: a worldwide survey**. New Delhi: International Commission on Irrigation and Drainage, 1976. 106p.

SHARMA, B. R.; MINHAS, P. S. Strategies for managing saline/alkali water for sustainable agricultural production in South Asia. **Agricultural Water Management**.v.78, p.136-151, 2005.

SHARMA, D. P.; RAO, K. V. G. K. Strategy for long term use of saline drainage water for irrigation in semi-arid regions. **Soil & Tillage Research**, v.48, p.287295,1998.

SILVA JÙNIOR, L. G. A.; GHEYI, H. R.; MEDEIROS, J. F. Chemical composition of crystalline water from the Brazilian Northeast. **Brazilian Journal of Agricultural and Environmental Engineering.** Campina Grande, PB. v.3, n.1 p. 11-17. 1999.

SILVA JÙNIO, G. de S. Biometric, Nutritional and Biochemical Parameters in Banana Cultivars and Somaclones Subjected to Salinity**. IFPE Journal of Science, Technology and Humanities**, v. 4, n. 1, 2012.

SILVA, D.; PRUSKI, F.F. Recursos **hidricos e desenvolvimento sustentâvel da agricultura**. Brasilia: MMA/ SBH/ABEAS, 1997. 252p.

SILVA, F. L. B. Interaction between salinity and bovine biofertilizer in the cultivation of beans - de-corda. **Revista Brasileira de Engenharia Agricola e Ambiental**, v. 15,

n. 04, p. 383-389, 2011.

SILVA, M.M.C.; MEDEIROS. J.F.; NEGREIROS, M.Z.; SOUSA, V.F. Melon fruit productivity under different levels of irrigation water salinity, with and without soil cover. **Horticultura Brasileira**, Brasilia, v.23, n.2, p.202-205, 2005.

SILVA, J. V. Physiologicalresponses of NaCl stressed cowpea plants grown in nutrient solution supplemented with CaCl2. **Brazilian Journal PlantPhysiology**, v.15, p.99-105, 2003.

SILVA, V. A. Physiological response of a Conilon coffee clone sensitive to water deficiency grafted onto a tolerant rootstock. **Pesquisa Agropecuâria Brasileira, Brasilia**, v.45, n.5, p.457-464, 2010.

SIVRITEPE N; SIVRITEPE HO; ERIS A. The effect of NaCl priming on salt tolerance in melon seedling grown under saline conditions. **Scientae Horticulturae**. p. 229-237. 2003.

SOARES, J. I.; COSTA, R. N. T.; SILVA, A. C.; GONDIM, R. S. Watermelon response function to water levels and nitrogen fertilization in the Curu Valley, CE. **Revista Brasileira de Engenharia Agricola Ambiental**, v. 6, n. 2, p. 219224, 2002.

SOUSA G. G.; MARINHO, A. B.; ALBUQUERQUE, A. H. P.; VIANA, T. V. A.; AZEVEDO, B. M. Biomass accumulation, content and extraction of micronutrients in corn plants irrigated with saline water. **Revista Agropecuâria Tècnica**, v. 31, n. 02, p. 1-10, 2010.

SOUSA, R. A.; LACERDA, C. F. DE; FILHO, J. A.; HERNANDEZ, F. F. Growth and mineral nutrition of string beans as a function of salinity and ionic composition of irrigation water. **Revista Brasileira de Ciências Agrârias**, v.2, p.75-82. 2007.

SOUSA, R. A. de. Effects of salinity and ionic composition of irrigation water on the development of string bean plants cv. Pitiùba. **Dissertation** (Master's) in Irrigation and Drainage - Center for Agricultural Sciences - University Federal do Cearà, 87 f., Fortaleza, 2006.

SOUSA, M. R. de. Behavior of bean plants (*Phaseolus vulgaris* L. CV Eriparza) subjected to different levels of salinity in irrigation water.**Dissertaçâo** (Master's Degree in Agricultural Engineering) Universidade Federal de Lavras, 94 p, Lavras, 1995.

SUMNER, M. E. Sodic soils: New perspectives. In: Naidu, R.; SUMNER, M. E.; RENGASAMY, P. (ed). Australian sodic soils: Distribution, properties and management. Adelaide: First National Conference and Workshop on Sodic Soils, p. 1-34, 1995.

SZABOLCS, I. Global overview of sustainable management of salt affected soils. **International Workshop on Integrated Soil Management for Sustainable use of Salt Affected Soils**. Philippines, 1995.Procceding... Philippines, Bureau of Soil and Water Management, p.19-38, 1995.

TABOSA, J.N. Behavior of forage sorghum cultivars in different agroecological environments in the states of Pernambuco and Alagoas**. Revista Brasileira de milho e sorgo**, Sete Lagoas, v.1, n.2., p.47-58, 2002.

TAYLOR, R. M.; YOUNG, E. F.; RIVERA, R. L. Salt tolerance in cultivars of grain sorghum.**Crop Science**. v.15, p. 735-735, 1975.

TAVORA F. J. A. F.; FERREIRA R.G.; HERNANDEZ F. F. Growth and water relations in guava plants submitted to salt stress with NaCl. **Revista Brasileira de Fruticultura.** 23(2): 441-446, 2001.

TERRY, N. & WALDRON, L. J. Salinity, photosynthesis and leaf growth. **California Agriculture**, 39: 38-9, 1984.

TIVELLI, S. W.; MENDES, F.; GOTO, R. Estimation of the leaf area of pepper (*Capsicum annum* L.) cv. Elisa conducted in a protected environment. **Horticultura Brasileira**, v.15, 1997. Supplements.

TOTAWAT, K.L.; MEHTA, A.K. Salt tolerance of maize and sorghum genotypes. Annals of Arid Zone, **Jodh Pur**, v.24, p.229-236, 1985.

VILLAS BOAS, R. L. **Nitrogen doses for peppers applied conventionally and through fertigation.** 2001. 123 f. Thesis (Livre Docência) - School of Agronomic

Sciences, Universidade Estadual Paulista, Botucatu, SP.

VAN GENUCHTEN, M. TH. A closed-form equation for predicting the hydraulic conductivity of unsaturated soils. **Soil Science Society of America Journal**, v.44, p.892-898, 1980.

MIX
Papier aus verantwortungsvollen Quellen
Paper from responsible sources
FSC® C105338

Printed by Books on Demand GmbH, Norderstedt / Germany